The Eye Book

The Eye Book

John Eden, M.D.

Illustrations by Laszlo Kubinyi

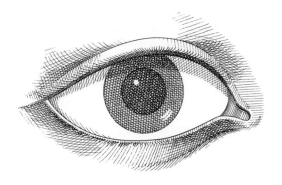

THE VIKING PRESS   NEW YORK

Copyright © John Eden, M.D., 1978

First published in 1978 by The Viking Press
625 Madison Avenue, New York, N.Y. 10022

Published simultaneously in Canada by
Penguin Books Canada Limited

LIBRARY OF CONGRESS CATALOGING IN PUBLICATION DATA
Eden, John.
The eye book.

Includes index.
1. Eye—Diseases and defects. 2. Eye—Care
and hygiene. I. Title.
RE51.E33   617.7   77–21578
ISBN 0–670–30289–9

Printed in the United States of America

Set in Videocomp Garamond

Contents

Acknowledgments

I wish to thank Barbara Kelman-Burgower, without whose clear thinking, energy and talent in the collaboration of this book it would not have been possible.

Additional thanks to my special friends whose encouragement and assistance were invaluable.

Introduction

Do you believe that reading in poor light can damage your eyes? That you can be so nearsighted that you are almost blind? That eating carrots will improve your eyesight? That contact lenses are dangerous because they can get lost in your eyes and even travel back inside your head? Or that contact lenses are particularly valuable because they keep your eyesight from getting worse? That you don't have to see an eye doctor regularly if you don't wear glasses? That children who are cross-eyed will grow out of the problem in the course of time? That eye make-up is bad for your eyes? That glasses can strengthen your eyes and, in time, make wearing them unnecessary? Or that wearing glasses can weaken your eyes and make you dependent on them? That cheap, nonprescription sunglasses can hurt your eyes? That we all get farsighted as we approach middle age and that's why we need reading glasses? That if you get glaucoma you will know it because of the pain and other symptoms you will feel? That proper eye care includes using eye drops and eye washes? That it makes no difference whether you go to an optometrist or an ophthalmologist for your eye care? That you will injure your eyes if you sit too close to a movie screen? That people who need glasses have less than healthy eyes?

If you believe any of these things, or the myriad other myths and misconceptions about the human eye and vision, you need this book.

In more than fifteen years of practicing ophthalmology, I have been asked tens of thousands of questions about the eyes and eye care, and I have, on a one-to-one basis, tried to replace the many incorrect notions my patients have with clear and useful facts about the eye. Because sight is the most precious of our senses, I believe we can all benefit from a fuller understanding of how our eyes work. This book grew out of my desire to bring that understanding to a wider audience and, particularly, to dispel the many myths and misconceptions commonly held about the eyes and eyesight.

In the pages that follow, I will explore with you the workings of the human eye, from birth to old age, and the most common eye problems, so you will understand what goes wrong and how it can be treated. I will also discuss steps you can take for preventive eye care, both with and without your doctor.

This book is not intended to make you an expert; rather, it has been written to help you to gain a better understanding of the eye in general so you can give your eyes, and those of your family, the most intelligent, economical, and anxiety-free care possible.

The Eye Book

1

How We See

MYTH: Babies can see at birth.

FACT: Although we are all born with two anatomically developed eyes, our vision at birth is by no means fully developed. Newborn babies see little more than the difference between light and dark. We all must learn to see, just as we must learn to talk, and this learning process takes place gradually between birth and age six.

MYTH: If you get something in your eye, it can get lost inside your head.

FACT: The exposed front part of the eyeball is separated from the back portion by the conjunctival membrane. This forms a closed cavity that makes it impossible for foreign bodies, such as particles of dust or contact lenses, to travel beyond the outer part of the eyeball to the inside of the head.

MYTH: Blue eyes are more delicate than brown eyes and are, therefore, more susceptible to irritation and other eye problems.

FACT: The only difference between blue and brown eyes is in the amount of pigment in the iris. Because blue eyes have less pigment, they may be somewhat more sensitive to light than brown eyes, but they are not otherwise more vulnerable to irritation or any other eye problems.

The sense of sight assures us of contact with the outside world perhaps more than any of our other senses. Our ability to see the shape, size, and color of things around us and to determine our relation to them is a gift of nature we tend to take for granted. Our eyes are remarkable instruments, perfectly designed to collect visual information. Along with the brain, to which they are connected by a specialized nerve network, our eyes provide us with the full range of vision on which we depend so heavily.

In the chapters that follow, I will talk about how our eyes work and how, at times, they do not. I will explain the ways in which many eye problems can be treated, sometimes by an eye doctor, other times at home. I will also try to explain why many widely held beliefs about the human eye and vision are incorrect, and will substitute for these myths solid facts that can help you, in partnership with your eye doctor, to take proper care of your eyes. But first I would like to take a look with you at the anatomy of the eye and the function of its various parts. I hope this will help you to understand how this amazing optical system enables you to see the world you live in.

Anatomy and Function of the Eyeball

The human eye is, of course, a dual organ—two eyes working together to transmit visual information to the brain. Although it is certainly possible to see with only one eye, it takes two normally functioning eyes to achieve normal vision.

Your eye is made up of numerous kinds of highly specialized cells, which perform different functions. It is equipped with muscular, fibrous connective, circulatory, and nervous systems of its own. Although they are similar to those systems that work throughout your body, they are designed to fill the special needs of the eyes.

The normal adult eyeball is an elliptical sphere, which

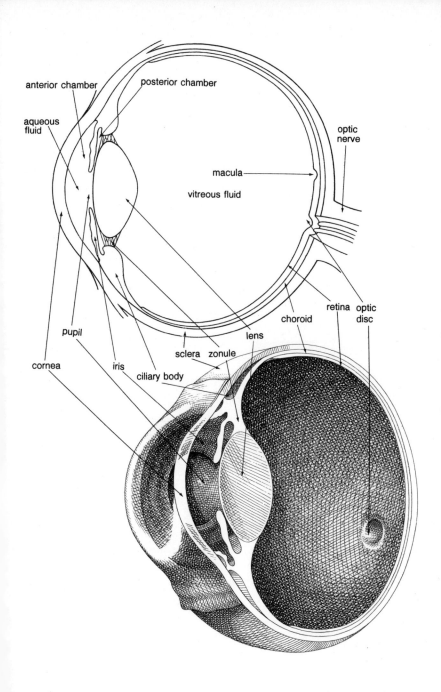

anterior chamber

posterior chamber

aqueous
fluid

optic
nerve

macula

vitreous fluid

pupil

cornea

iris

sclera

ciliary body

zonule

lens

choroid

retina

optic
disc

3

means it is more egg-shaped than perfectly round. It has three distinct concentric tissue layers. The first serves to protect your eye's delicate internal structures, and it consists of the *sclera**—the opaque white of the eye—and the *cornea*—the transparent layer that lies in front of the pupil and iris.

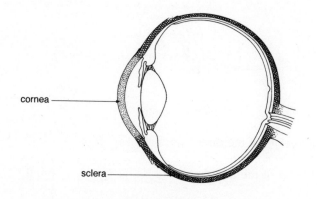

The protective layer of the eyeball.

The sclera covers about five-sixths of the surface of the eyeball. It is interrupted only by the cornea in front and the optic nerve, which enters the eyeball at the back. Although not much thicker than the page you are reading, the cornea and sclera are composed of extremely tough tissues. I will not say it is impossible to pierce them, but it takes a very sharp object traveling at high speed to do it.

A thin membrane called the *conjunctiva,* which is not technically a part of the eyeball, separates the exposed front and unexposed back portions of the eyeball. It covers the front part of the sclera and then laps over and continues forward onto the inner surface of the upper and lower eyelids. The conjunctiva thus closes off the back part of the eyeball, making it impossible for anything to get lost in your eye or travel back into your head.

*All words defined in The Layman's Eye Dictionary, beginning on page 183, will appear in italics the first time they are used in the text.

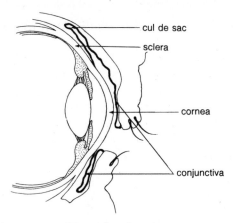

The conjunctiva.

The second of the three layers is called the *uveal tract,* and its main functions are circulatory and muscular. The uveal tract is made up of the iris, the ciliary body, and the choroid.

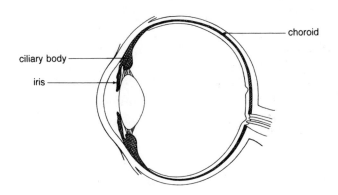

The uveal tract.

The *iris* is the round colored part of the eye that surrounds the pupil, and it is responsible for what we call the color of our eyes. The main function of the iris is to permit more or less light to enter your eye. The pupil itself is simply the hole surrounded by the iris and it is through this hole that light passes into your eye. The involuntary muscles of the iris respond primarily to the stimulus of light, constricting to make a smaller hole when light is bright and dilating to make a larger hole when light is dimmer. This action is like that of the iris diaphragm in a camera. But please don't take this analogy too literally. The human iris is not a mechanical device whose opening can be varied whenever you decide to do it. The action is involuntary. The muscles of your iris do not snap nearly shut when light is very bright and zoom open when light dims, but they constantly adjust and readjust to the level of light, remaining stationary only when the level of light stays the same. Your iris responds to all changes in light, no matter how subtle, and the adjustments are often extremely minute. The change from a darkened movie house to bright daylight is a dramatic one, but it is by no means the only sort of adjustment your iris makes.

The *ciliary body* lies between the iris and the choroid, and its function is also primarily muscular. It is connected to the lens by a ligament-like tissue called the *zonule,* from which the *lens* is suspended like a person in a hammock. The muscles of the ciliary body contract or relax to alter the shape of the lens, which allows your eye to focus for near vision, refocus to see at the distance, and then back again to see nearby objects clearly. When I speak of the *focusing muscles* of the eye, I am referring to the muscles of the ciliary body. When I speak of bringing an object into focus, I mean the bending of the lens

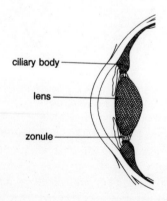

ciliary body

lens

zonule

The structures that allow your eye to focus for near vision.

by these muscles to change the shape of the surface through which light passes. The ciliary body also has a secretory function—it is from here that the *aqueous fluid* originates.

Behind the ciliary body is the *choroid,* the main circulatory layer of the eye, through which blood is carried to nourish the various parts of the eye. This is not your eyeball's only blood supply; your retina, for example, has its own circulatory system. From the choroid, increasingly tiny arteries branch out to those portions of the eye that require blood for their metabolism, and then the veins return to the choroid blood laden with waste in the form of carbon dioxide. Some parts of your eye cannot be nourished in this way since the presence of blood vessels would interfere with their optical function, but nature has compensated for this by providing other ways of getting oxygen to those tissues and carrying away waste.

The innermost layer of the eye is the *retina,* an extremely thin sheet of specialized nerve tissue made up of ten distinct cell layers, each of which performs a specific part of the task of receiving visual images and transmitting them, via the optic nerve, to the brain. To go back to our camera analogy, the retina is like the film on which images are focused and recorded, but of course the retina is more complicated and can function more effectively than film. Your retina receives and passes along to your brain a very complex visual message,

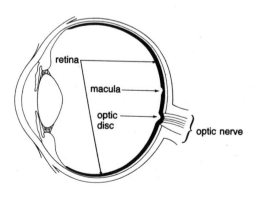

The retina.

which includes size, shape, dimension, position in space, relative distance, and color.

The key central area of the retina, located slightly to the outer side of the eyeball, is called the *macula.* This tiny area, which represents only a small part of the total retina, is its most vital part. It is responsible for your sharp central vision, and it is what permits normal 20/20 vision; the rest of the retina receives peripheral, or side vision, and delivers an image that is not so sharp as that coming from the macula.

Although blood vessels crisscross almost the entire retinal surface, the macula cannot be fed in this manner, since the highly sensitive receptor cells would be obscured by the blood vessels. Small capillaries feed into the edge of the macula, but the exchange of oxygen and carbon dioxide in the center takes place by absorption through cell walls. Nature's solution to the problem works quite well as long as nothing interferes with the delicate process, but it does make the macula more vulnerable to damage than the rest of the retina.

In addition to its receptor properties, the retina is able to adapt to light and dark. The iris performs the task of admitting or excluding light from the interior of your eye, but in addition to this, certain cells of the retina—the familiar *rods* and *cones*—undergo photochemical changes to enable you to see in various light levels. When you go from the daylight outdoors into a more dimly lit room, the rods in your retina are activated and the cones deactivated to adjust to the lower level of light; when you return to the sunlight, the cones are again activated and the rods function less so you can adjust to the brighter light. It takes a bit of time for your retina to adjust to the light change—an hour for complete *light* or *dark adaptation,* though you will see well in much less time—which is why when you go inside on a sunny day the room often seems quite dark for a while until your eyes adjust to the new light level. The cones are also responsible for your ability to perceive colors.

All the visual information collected and recorded by your eye is transmitted to the brain by the optic nerve, which enters the eye at the back of the retina. Because there is no retinal tissue at that point, this results in a *blind spot,* an area that cannot receive visual messages. Your eye doctor can locate and measure your blind spot by covering one eye at a time and performing a special test, but under normal conditions you do not notice your blind spot because the area it cannot see is seen by your other eye. However, even if you use only one eye, your blind spot is not a practical reality since it is so small.

Now that we have examined the tissue layers of the eyeball, let's look inside. What gives your eye shape, and, more important, what does light pass through in order to bring all that visual information to the retina? Imagine a ray of light passing directly from the object you see to the retina. This imaginary line is called the *visual axis,* and it must run through a perfectly clear series of structures so that no light is lost and no distortion or obstruction of the image takes place. These clear structures are called the *optical media,* and they are, from the front to the back of the eye, the cornea, the aqueous fluid, the lens, and the vitreous fluid.

Like the macula, the cornea must not be obstructed by blood vessels, which would interfere with its optical clarity. The cornea's metabolism is taken care of by tears on the outside and aqueous fluid on the inside. Both these liquids can carry oxygen and carbon dioxide, just as blood does, but they are clear, so light can pass through them. Tears are secreted onto the surface of the eyeball by *lacrimal glands,* which are

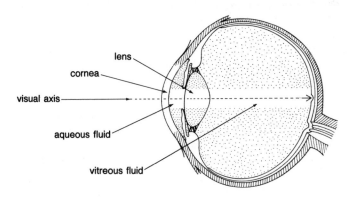

The visual axis passes through the clear optical media.

located inside the bony eye socket, and drained away through small passages located in the eyelids. Aqueous fluid is extracted from the blood that the choroid brings to the eye. It flows from the area between the lens and the iris (the *posterior chamber*) into the area between the iris and the cornea (the *anterior chamber*). The waste-laden fluid is then reabsorbed into the blood. This dynamic circulation means that the aqueous fluid is constantly secreted into and drained away from the eye.

In addition to its protective function, your cornea acts as a light-bending *(refractive)* surface in the visual axis. It is the first curved structure light hits, and it bends the angle of the light rays inward, narrowing the beam as it enters the eye and travels through the aqueous fluid to your lens, which bends it further.

Although you can understand much about the lens of the eye by comparing it with various glass optical lenses, nature's lens is really much more complicated and effective. It is not a rigid piece of glass, but a flexible unit that can alter its shape to permit us to see objects clearly, whether they are at the distance or near at hand. Like the cornea, the lens must be perfectly transparent, so there is no vascular system in it. Instead it receives its main nourishment from the aqueous fluid.

The lens grows throughout life. As in a tree, new cell layers are laid down on the outer surface in a continual concentric pattern. Although there is a slight increase in size each time a layer is added, the layers become compressed with age and tend to become less elastic and, later in life, less clear.

Behind the lens, the bulk of the eyeball is filled up with another optically clear mass called the *vitreous fluid.* Unlike the aqueous, the vitreous is a stable, gel-like mass that does not circulate and cannot be manufactured by the eyeball. Its function is mostly to maintain the shape and resilience of the eyeball and to provide a clear medium through which light can pass to the retina after having been bent by the lens.

Each eyeball is suspended inside a cone-shaped skeletal socket called the *bony orbit,* which has a hole in the back through which the optic nerve, the main eye blood vessels, and other nerves that serve the eye run. The orbit protects the eyeball since the upper and lower ridges (your brow and cheekbone) are the first surfaces to receive the impact from a blow to the eye. Fatty and fibrous tissue surround the eyeball to cushion it within the orbit.

The upper and lower eyelids also protect your eye, and your eyelashes are an effective barrier against dust and other airborne bodies. In addition, your eyelids are lined with conjunctival membrane and contain the passages through which tears enter and drain from your eye to keep the exposed part of the eyeball nourished, lubricated, and smooth.

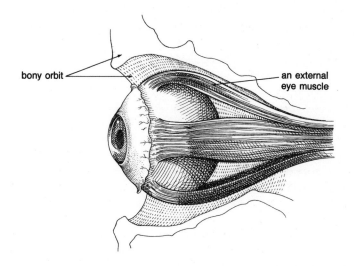

Eyeball in socket, showing three of the external eye muscles.

The inner eye muscles of the ciliary body are responsible for bending the lens, but there are also external muscles that control eye movement. There are six main muscles in each eye, paired to allow you to move your eyes 360 degrees. The muscles are attached around the side of the eyeball, behind the conjunctival barrier, and they are connected to the back of the orbit.

The Visual Process

Now that I have described the various components of the human eye and their functions, let's consider how they all work to make vision possible. The camera resembles a simplified eye, but once the image has been recorded on film, the process is finished. A lab must develop the film if anything is to be seen. In the human eye the visual process has really just begun when rays of light reach the retina.

The interpretation of visual data by the brain is a significant part of seeing, but many people do not understand that seeing is a learned process. It is not enough to have two healthy eyes with all their working parts in order; we must also learn how to interpret what they deliver to the brain.

We do not see at birth. We are not blind, but we simply do not have the learned capacity to "read" the information brought to our brain by our eyes. Learning to see takes place gradually, so that we cannot say, for example, that at such-and-such an age you first learn to distinguish colors or at another age you are able to tell that one object is in front of another. From the moment you are born, you begin to learn what the messages mean as your eyes develop and as you apply experience to what you perceive. You gradually come to see clearer and more refined images, and bit by bit your distance and near vision become more acute. Soon you develop three-dimensional vision and begin to distinguish colors more clearly.

In general, the learning period is thought to be from birth to age six. You have probably observed that a child can easily learn a new language and speak without an accent in early childhood, but that language skills become more difficult to pick up as the child grows older. In the same way, we learn to perceive size, color, shape, distance, spatial relations, and three dimensions before we reach the age of six. If there are any abnormalities that interfere with this process and go uncorrected until we have passed that age, it is virtually impossible to develop normal vision.

How, then, do we see? How do our eyes work together to collect light rays and bring them to the retina, and how is this focused image carried from the retina to the brain? Let's trace what happens in an adult eyeball when it views an object—a flower, for example.

In order to see the flower, there must be light. You do not actually see objects; you see light reflected off objects, which is why you cannot see anything at all in total darkness. The less light there is, the less clearly you will see. Your retina will adapt and your iris will dilate or constrict according to the level of available light, but there must be at least some light present.

The air between the lighted flower and the front surface of the eye must be clear. You know that it is harder to see when it is foggy or when you look through a dirty windowpane, and that it is impossible to see at all if there is a solid object blocking the light directly in front of the thing you want to see.

If all goes well and the light thrown off from the flower has traveled through unobstructed space, light rays will reach your cornea, where they are bent inward by its curved surface and transmitted through the aqueous fluid and onto the lens. At this point the lens bends the light rays farther, and they then travel through the clear vitreous fluid and finally end their journey as a focused image on the surface of the retina. Because the retina is not optically clear, the light cannot travel any farther.

At this point the image of the flower is upside down and flat. There is no mechanism within the eye to reverse it so it is presented right side up—no mirrors or prisms. The righting of the image is accomplished in the brain and is the result of a learned process.

Seeing objects as solid and three-dimensional is also something you must learn. Three-dimensional vision cannot exist if only one eye is doing the seeing or if your eyes are out of alignment and looking at different things. But even if your two

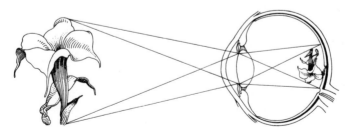

A focused image arrives at the retina after traveling as rays of light through space and the clear optical media of the eye.

eyes are aligned, a three-dimensional picture cannot be seen unless and until your brain has learned how to interpret the information it gets.

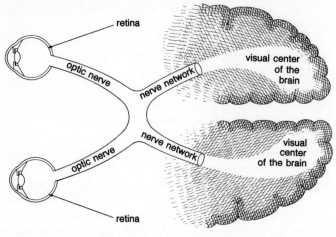

The visual pathway to the brain.

Visual data is transmitted by the *optic nerve* along a nerve network that travels to the visual center of the brain. The message is in a sort of code that your brain has learned, and it contains information about the size, shape, and color of the objects, and it reports as well which part of the retina received the image—a kind of return address. For example, an image received by the macula and recorded by the highly specialized macular cells will be received by the brain as a clear object viewed straight on. An image received by a peripheral part of the retina will arrive at the brain with data about its retinal origin, and the brain will "read" it as an object off to the left or right or wherever.

Our eyes and brain work together in this fashion to give us visual knowledge of things around us. Although it is vital to sight that we be equipped with two structurally normal and healthy eyes, they alone will not permit us to see. While the human eye is a perfectly designed collector of visual data, the human brain is the perfect decoder, and the partnership of the two is what gives us our highly sophisticated visual system.

2

The Optical Errors:
Nearsightedness, Farsightedness,
Astigmatism, and Presbyopia

MYTH: You get more farsighted or less nearsighted as you approach middle age.
FACT: Both farsightedness and nearsightedness tend to stabilize at about age twenty-one. The condition people are referring to when they make the above claim is called presbyopia, which is a natural process of aging that we all go through. Presbyopia is simply the loss of our ability to focus on near objects. It is not the same as farsightedness and has no effect on our ability to see distant objects clearly.

MYTH: Eye exercises will improve nearsightedness, farsightedness, and other eye problems.
FACT: We exercise our eyes every time we use them. Additional exercises serve no purpose whatever. All optical errors, including nearsightedness and farsightedness, are caused by factors inside the eye. No type of eye exercise has any effect on its internal optics.

MYTH: If you do not see well at the distance, your eyes are weak.

FACT: Poor distance vision can be caused by one of several optical errors, none of which has anything to do with eye weakness or eye disease. The implication of the word *weak* is that the eye has some muscle problem or is diseased or prone to disease. This is absolutely false. Optical errors do not mean that your eyes are unhealthy in any way.

Nearsightedness, farsightedness, astigmatism, and *presbyopia* are the most common problems we have with our eyes, and they are the overwhelming reasons we wear glasses or contact lenses. But there is great confusion and misunderstanding concerning these problems.

The first and perhaps most important thing I want to say about these four conditions is that they are *optical errors,* not eye diseases. These errors can be corrected and normal vision can be achieved. Eyes that are nearsighted, farsighted, astigmatic, or presbyopic are not unhealthy, nor are they more susceptible to eye disease than eyes without optical error. And by no means is blindness the extreme of any of the four. The correction for optical errors is corrective lenses—either glasses or contact lenses—which will allow perfect or near-perfect vision when they are worn. In no case can glasses, contact lenses, or anything else *cure* these conditions and make it possible for the patient to see well without corrective lenses. This is because nothing—including eye exercises, surgery, or specially designed contact lenses—can alter the internal optics of the eye. Attempts to "cure" optical errors are regarded by the medical profession as a waste of time and money on the part of the patient.

What Is Normal Vision?

If optical errors are not eye diseases, what are they? They are specific defects in the internal optics of the eye that prevent normal vision. Okay, but what is normal vision? Isn't that a pretty vague term for something as important as seeing well? Actually, *normal vision* is a very specific term that describes the ability to see 20/20 with both eyes and to use both eyes together to deliver a three-dimensional visual message to the brain.

Just about everyone has heard the term *20/20* used as a measurement of vision, but few people know what it means and usually assume it describes the two eyes in relation to each other. In fact, each eye is measured separately for *20/20 vision;* the ratio refers to the difference between the eye being measured and a theoretical normal-seeing eye. For example, if you have 20/20 vision in one eye, that means that your eye sees at 20 feet what a normal eye can see at 20 feet. In other words, your eye has normal vision. If you have 20/400 vision in one eye, that eye has to be as close as 20 feet to see what a normal eye can see at 400 feet. That eye has considerably less than normal vision. A 20/15 eye can see at 20 feet what a normal eye must be within 15 feet to see, and is better than normal. You can have the same vision in both eyes, of course, but you can also have one eye that is 20/20 and one that is 20/100 or whatever.

The man with 20/20 vision can see the vision chart clearly from 50 yards away. The man with 20/50 vision must move to the 20-yard line to see it as clearly.

I am often asked by my patients what the worst possible vision is. There is no limit; one can have 20/6000 or even worse.

Because the United States is in the process of converting to the metric system, more and more doctors are beginning to use 6/6 as the expression of normal vision. The idea is, of course, the same, but meters rather than feet are the unit of measurement.

Vision is measured both at the distance and close up. Eye doctors take the measurement for *distance vision* at twenty feet, and for *near vision* at a normal reading distance, or about fourteen inches. Near vision is what you use for reading, sewing, eating, drawing, or any other situation in which what you are looking at is within arm's reach. In practical terms, distance vision is anything beyond that. I have heard many people say they think of distance vision as looking at things really far away, but such things as looking at television or at people just across a room involve distance vision too. Middle distance, which is neither near nor far, is not part of the ophthalmological measurement, but it does play a part in your viewing experience, particularly as you get older. *Middle-distance* viewing situations include such activities as reading piano music, cooking, and playing such card games as poker. (The cards you hold require near vision; the cards across the table are beyond arm's reach but not really far away.)

Many people tend to assume that the glasses they have to correct a distance vision problem are to be worn only for viewing faraway objects. But for people with optical errors, the distance for which they require correction begins at the point where they start to experience difficulty seeing clearly.

What Are the Optical Errors?

If you have 20/20 vision at the distance and close up, you probably do not have an optical error. But let's examine what the specific optical errors are and how they affect your vision. In the discussion that follows, I will be talking about the characteristics of *uncorrected* vision, but you should keep in mind that all these optical errors are correctable.

Nearsightedness

Myopia is the technical term for nearsightedness. Although I will use the common name when I talk about this optical error, actually it is a bit of a misnomer. First of all, nearsightedness sounds like an advantage rather than a vision problem. Although it is true that people who are nearsighted see better close up than they do at the distance, they do not see better than 20/20 close up. Very nearsighted people, for example, cannot see well either at a normal reading distance or far away. These people often have to hold things very close to their eyes to be able to see clearly at all without correction. Less nearsighted people may be able to see quite well at conversational distances, but have trouble seeing clearly across a room. The farther away an object is, the more blurred it looks to a nearsighted person. The more nearsighted a person is, the closer to the eyes that blur begins.

Let's review for a minute what normally happens when we look at an object. Light rays travel from the object to our eyes. They enter the eye through the cornea, which bends the rays slightly inward. They then travel through the lens, which narrows them even more. The rays continue to narrow until they converge and are in focus when they reach the retina. The retina receives the light rays and sends a message of the visual image to the brain. But the retina can send to the brain only as good an image as it receives. If the image is in sharp focus, a sharp, clear picture will be perceived. If it is not in focus, a blurred picture will be seen.

In the nearsighted eye, the light rays converge and become focused before they get to the retina. From that point, they start to diverge and go out of focus. This means that by the time light rays reach the retina, they are no longer focused. The retina sends that unfocused message to the brain, and a blurred picture is perceived. How blurred the picture is de-

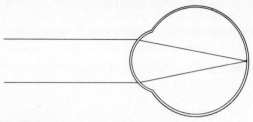

No optical error (emmetropia).
The image comes to focus on the retina.

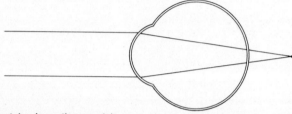

Farsightedness (hyperopia).
The image comes to focus (theoretically) behind the retina.

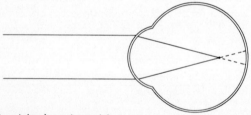

Nearsightedness (myopia).
The image comes to focus in front of the retina.

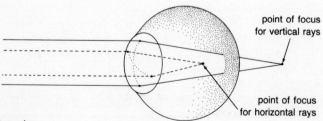

point of focus
for vertical rays

point of focus
for horizontal rays

Astigmatism.
The image comes to focus at two separate points.

pends on how far in front of the retina the point of focus is. It can be pretty close, which results in only a slight blur, or it can be quite far in front, which results in a very great blur.

A number of different factors can cause nearsightedness. The most common is an eyeball that is longer than normal and requires the light to travel a greater distance after it has entered the eye. But before you run to the mirror to see if your eyeball is too long, let me explain that I'm referring to the distance from the front to the back of the eyeball, which is something you cannot see. It has nothing to do with whether your eyes look prominent or bulging. Besides, the eyeball is small enough that the tiniest variation in shape is enough to make a difference. Another cause of nearsightedness can be an eye whose light-bending *(refractive)* ability is more than is required to get a focused image to the retina.

I do not tell my patients what has caused their nearsightedness, mostly because I usually cannot determine the cause myself. Also, it is not really important since I can do nothing to change the underlying cause. What I can do is prescribe corrective lenses, which will do the job the nearsighted eye cannot do by itself. No doctor can alter your intrinsic optical system, shorten your eyeball in any way, or give you eye exercises that will make your eyes stronger.

Nearsighted people can achieve slightly greater clarity by squinting. This is somewhat like closing down the aperture of a camera to increase depth of focus, but it certainly does not work as well as corrective lenses, and it can cause eyestrain and headaches. It is also not very pretty.

Nearsightedness usually begins to be noticeable during the growth period, which ends around age twenty or twenty-one. During that time, the statistical probability is that the degree of nearsightedness will increase until the end of that period and then stabilize. Naturally, people are not statistics, and in any individual case the degree of nearsightedness may remain the same throughout the growth period. Or it can even improve, although this is quite uncommon. Nearsightedness can, of course, also get worse or better after an individual has reached adulthood.

The usual tendency for nearsightedness to get worse during

the growth period has given rise to the term *progressive myopia*. I do not like to use it, particularly because it suggests a disease that keeps getting worse. Myopia is not a disease at all, and although the condition can worsen as you and your eyes grow, it usually stops "progressing" by the age of twenty-one. After that point, you will probably not get any more nearsighted, but if you do, don't be afraid that you will become so nearsighted that you are blind. Many patients express this fear to me, but it is completely groundless. As long as your vision is correctable to 20/20, the only thing you have to worry about is the nuisance of being dependent on corrective lenses.

Farsightedness

Hyperopia, or farsightedness, is another common optical error. And, as with nearsightedness, its common name is a confusing way of referring to the condition. For one thing, the effect it has on vision is not the opposite of nearsightedness—a farsighted person does not see well at the distance and badly close up as opposed to seeing well close up and badly at the distance. The practical effect that farsightedness has on vision is not so simple to understand as that of nearsightedness because, in addition to the degree of the error, the age of the individual is an important factor. A young farsighted person may have no problem seeing well both up close and at the distance; an older farsighted person may see equally poorly near and far.

Let's look again inside the eye to try to understand what this is all about. In a farsighted eye the light rays are not yet in focus when they reach the retina. Some explanations of farsightedness say that the light rays come to focus *behind* the retina. I prefer not to describe it this way because it implies that the rays could pass through the retina and beyond to some mysterious place where they come into focus. Light stops at the retina, and the state of focus at that point is the state of the picture that gets sent to the brain. If light could go farther, it would focus at a certain hypothetical point, and the farther that point was beyond the retina, the greater the farsightedness.

But the important thing to remember is that the rays of light are *not* in focus when they arrive at the retina, and this is what causes the blurred vision of farsightedness.

It may be easier to understand and distinguish between nearsightedness and farsightedness if you think about them in the following way: the point of focus in nearsightedness is on the *near* side of the retina; the point of focus in farsightedness is on the (theoretical) *far* side of the retina. I hope this will help avoid the confusion that comes from considering the problem factor to be the nearness or farness of the object being looked at.

The tricky thing about all this is that, whereas nearsighted people can do nothing to improve their vision without corrective lenses, young farsighted people can. This is what makes the age of the farsighted person an important influence on what vision problems are experienced.

The near-focusing muscles, which are located inside the eye

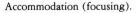

Accommodation (focusing).

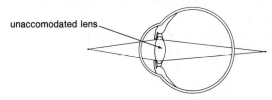

An image brought to focus behind
the retina in an unaccommodated (unfocused) eye.

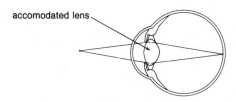

The same eye with an image brought to focus on the retina
through accommodation (focusing).

The eye can also use the normal process of accommodation to correct farsightedness. Viewing a distant object, the near-focusing muscles are used to bring the image forward to the retina.

and normally used to bend the lens to bring near objects into focus, can also be used by a farsighted eye to bring distant objects into focus. This is not a voluntary action that can be accomplished when you want to, like raising your arm or bending your leg. It happens without your noticing it. If you correct your farsightedness with your near-focusing muscles, you may have 20/20 vision and no problem seeing at the distance. But because you are using muscles intended to be used only when you focus on near objects, you may experience eyestrain and headaches. This is because the near-focusing muscles are being used all the time, not just for near viewing as nature intended. The resultant state of constant spasm can cause secondary symptoms.

But if you have 20/20 vision, how will your eye doctor know that this is a problem? One clue will be your complaints of headaches and eyestrain, but, even without this clue, we have a foolproof way of uncovering farsightedness, whether or not correction is made with the near-focusing muscles. The special eye drops eye doctors use when they examine your eyes contain a substance to paralyze the near-focusing muscles temporarily. It is then possible to measure your optical error without interference from those muscles.

Besides headaches and eyestrain, there is another catch to using the near-focusing muscles to correct farsightedness. It is a tool available only to the young. The lens-bending ability of these muscles decreases throughout life and is totally lost by mid-fifty to sixty years of age. As a farsighted person gets older, he or she will be able to use those muscles less and less, and finally they will be of no use at all, either for near or far vision. It is because of this that farsighted people may not realize they are farsighted until they approach middle age. This does not mean, as many people suppose, that they have suddenly developed "progressive hyperopia." They are simply gradually losing their innate ability to correct their farsightedness and are thus noticing vision problems for the first time.

Some degree of farsightedness is normal in growing children because the eyeball has not finished growing to adult size and shape. Ideally, the farsightedness decreases as the eyeball grows. (And, of course, growing children usually do not need to wear glasses for the slight farsightedness normal in childhood because they can use their very strong focusing muscles to overcome the optical error.) If, however, an individual reaches twenty-one with some farsightedness, he or she will probably remain that farsighted for the rest of his or her life. Farsightedness does not generally get worse after the age of twenty-one, for, like nearsightedness, it tends to stabilize when the eyeball stops growing. Remember, however, I am talking about statistical probabilities, so any one individual's experience might be different.

Farsightedness can be caused by an eyeball that is slightly shorter than normal or by an eye whose light-bending ability is less than required to deliver a focused image to the retina. Like nearsightedness, the root cause is usually impossible to determine and of academic interest only, since the solution is correction, not cure. If the error requires correction, the only effective approach is corrective lenses. These will provide good vision and, for younger farsighted people, relieve the symptoms of eyestrain and headache by taking the burden off the near-focusing muscles.

Astigmatism

Astigmatism is an optical error that is caused by a variation in the shape of the cornea. Rays of light pass through the cornea on two planes—vertical and horizontal. The normal cornea is shaped like a perfect hemisphere, with the same degree of slope on each plane. This makes the rays of light come together at a single point on their way to the retina. In the astigmatic eye one corneal plane is steeper than the other, so the rays of light do not come together at the same point. Instead, two separate images are focused at two separate points along the visual axis. One point could be on the retina and the other in front, or one could be on the retina and the other (theoretically) behind it, or they could both be in front or both behind the retina. The type of astigmatism is named for the location of the points of focus: *nearsighted astigmatism, farsighted astigmatism,* or *mixed astigmatism.*

The fact that there are two points of focus does not mean that astigmatic people see double. The effect is simply a blur, which is experienced both for distance and near vision. The greater the distance between the two points of focus, the more astigmatic that person is and the greater is the blur he perceives. Astigmatic people may also suffer from eyestrain.

Astigmatism tends to develop during the growth period. It is most likely that the general degree of astigmatism you have at age twenty-one will stay with you the rest of your life. It is certainly not impossible, however, for the degree of astigmatism to change, for better or worse, in any individual case.

Unlike farsightedness, astigmatism cannot be corrected with the near-focusing muscles, since the error is in the shape of the cornea, which cannot be changed. If the astigmatism or discomfort from eyestrain is bad enough, corrective lenses are the only possible solution.

Presbyopia

Presbyopia is the fourth optical error, and it is the only one of the four that we are all certain to acquire at some point in our lives. Unlike the other three, which are really structural defects, presbyopia is part of the natural process of aging. And, also unlike them, presbyopia is a visual problem experienced only when looking at objects close at hand.

In simple terms, presbyopia is the loss of the ability to use the near-focusing muscles. It is what makes people need reading glasses when they reach their mid-forties. And, of course, it is the condition I was talking about in the case of older farsighted people who are no longer able to correct their optical error on their own.

We know that the near-focusing muscles bend the lens to enable us to focus on near objects. As we grow older, the lens gets more rigid and more difficult to bend. This is a process that goes on throughout life and culminates at about age fifty-five to sixty, at which time all lens-bending ability is lost. This is a predictable phenomenon—so predictable, in fact, that your eye doctor can determine your age with great accuracy simply by taking into account your optical error (if any) and then observing how close up you can focus. So be warned: even if you can fool your friends, you can't hide your age from your eye doctor.

Young children can focus on an object held very close to their eyes. The point at which it is possible to focus gradually recedes as we grow older, but it presents no practical problem until it gets beyond comfortable reading range. This happens in the mid-forties and is the point at which people start joking that their arms are not long enough to hold a book.

Presbyopia is not the same thing as farsightedness. An entirely different thing is going on inside the eye, and blurred vision is present only when the eye views nearby objects, while distance vision is not affected by presbyopia at all. In fact, one can be presbyopic and farsighted (or nearsighted or astigmatic) at the same time. If you have an optical error, you will still have it in addition to the symptoms of presbyopia beginning in your mid-forties.

You have probably heard people say that if you are near-sighted when you are young, you will start to get farsighted, and, at about forty, you will reach the mid-point and have perfect vision. This notion is based upon myths piled on mis-understandings. Let's try to take it apart, myth by myth, so you do not make this same mistake. First of all, when people say farsighted in this instance, they really mean presbyopic. They are calling the inability to focus on close objects *farsightedness* since they think it is the opposite of nearsightedness, which they mistakenly think of as the ability to see well close up. The nearsighted person has difficulty seeing at the distance because of the optical error that brings images into focus before they reach the retina. When the decline of this person's focusing ability becomes noticeable, he or she also begins to have problems seeing near objects. The point at which he sees best (which is really least worst, so to speak) is in the middle distance—far enough away so the near-focusing muscles do not have to work beyond their capabilities, but not so far away that his nearsightedness starts giving him problems. However, it is certainly not true that his optical error is improving.

And while we are debunking myths, I would like to make it clear that farsighted people who use their near-focusing muscles to correct their distance error are not hastening the loss of their focusing ability. Age, not amount of use, is what is responsible for the loss, and nothing can make it happen sooner or later than nature intended.

The correction for presbyopia is reading glasses. These need to be worn for any kind of close viewing situation. If you have no other optical error, reading glasses are all you need. If you wear glasses to correct another optical error, you may need to wear bifocals or have two separate pairs of glasses, one for seeing at the distance and one for near objects.

Between the ages of forty and sixty, when your focusing ability is noticeably declining, frequent changes of your prescription are required. By mid-fifty or sixty years of age, all focusing ability is lost and your presbyopia cannot get worse.

It should now be clear that the four eye problems we have been discussing are imperfections of the optical system, not diseases. The treatments are not cures, but they are corrective measures. I'd like to underline this point once more by saying that eyes with optical defects are not diseased eyes. There is no reason for you to fear blindness as the end result of an optical error. The important thing to remember is that your eye doctor can correct your vision. This will mean that you must wear glasses or contact lenses to see well, but your dependence on them cannot be considered more than a nuisance, however great you may feel it to be.

3

Eyeglasses

MYTH: Wearing glasses will weaken your eyes or, conversely, wearing glasses will strengthen your eyes.

FACT: Both these contradictory notions are equally false. Glasses correct optical errors when they are worn, but they have no effect—good or bad—on the optical system within the eye and cannot, therefore, make the eyes weaker or stronger.

The notion that glasses do the work instead of your eyes is probably at the bottom of each of these myths. Some people assume that "vacationing" eyes are getting stronger because they do not have to work so hard, while others believe that the eyes get weaker from lack of exercise. Both ideas are absurd. When you wear glasses your eyes are working as hard as they can, just as they do when you do not wear glasses. The only difference is that you see better when glasses do what your eyes cannot do alone.

MYTH: If you have 20/20 vision, it definitely means you do not need glasses.

FACT: Oddly enough, this is not true. For example, young people who are very farsighted can use their strong near-focusing muscles to see clearly at the distance and thus correct their farsightedness. A simple test of their vision will yield a reading of 20/20, but the constant use of these muscles for other than their intended function can cause eye fatigue, eye-strain, and headaches, especially while they are reading. Glasses will alleviate the symptoms and allow 20/20 vision without taxing those muscles.

MYTH: You can damage your eyes if you wear somebody else's glasses.

FACT: It certainly is not a good idea to use a prescription that is not intended for your eyes, but you cannot inflict any damage to your eyes if you do. Somebody else's glasses could give you a headache, make you dizzy, and cause eyestrain, and you probably would not see very clearly, so using them is not recommended.

MYTH: Tinted glasses are bad for your eyes and should not be worn indoors.

FACT: Slightly or heavily tinted glasses can be worn if you enjoy the cosmetic effect or want to shield your eyes from glare or bright sunlight, but they are of no medical benefit or disadvantage. You can even wear sunglasses indoors where light levels are low and not endanger your sight in any way; it is not recommended since it makes it more difficult to see, but that is simply because you are reducing the amount of light by which to see, not because you are endangering your eyes.

MYTH: If you read a lot or do a lot of close work, you will ruin your eyes and make yourself need glasses.

FACT: Optical errors are never caused by reading or by any other heavy visual demand. The causes of these errors are within the eye, and you cannot inflict them on yourself in any way. People who read a lot may be more aware of optical errors and symptoms of headaches and eyestrain, which will give them a greater motivation to correct their problem with glasses or contact lenses. But the reading is certainly not causing the optical error.

MYTH: Cheap sunglasses are bad for your eyes.

FACT: Good sunglasses are recommended because they have sturdier frames and better-quality lenses, which create less distortion than cheap sunglass lenses. Good sunglasses are a better buy because they will last longer, but cheap sunglasses will in no way injure your eyes.

Eyeglasses have been around for at least a thousand years. Indeed, there is evidence that the ancient Egyptians and Assyrians used some sort of magnifying lenses to improve their vision. No single individual can be identified as the inventor of eyeglasses; by the tenth century glasses turned up simultaneously in such distant and unrelated places as Europe and China. When Marco Polo made his journey to the East, he observed that the Chinese used specially ground lenses to read fine print and unground colored crystals to protect their eyes from the sun's glare. In Europe, during the Middle Ages, monks were just about the only beneficiaries of improved sight through glasses. As the literate elite in an age of mass illiteracy, these men were among the few who needed to have perfect vision. This does not mean that kings and peasants, farmers and crusaders, courtiers and serving maids did not have eye problems. It simply means that their visual needs were not great enough to justify the use of glasses—their vision problems were not enough of a nuisance to interfere with the way they spent their time. And although the ways we spend our time today are quite different, the same principle still applies.

Eyeglasses will help correct any of the four optical errors we discussed in the last chapter and will relieve secondary symptoms of these errors. But they do this when, and only when, they are worn. Whether or not a particular individual wears glasses depends not only on how great his optical error is, but also on how well he wants to see and how much demand he puts on his eyes in various viewing situations. The monk who spent hours reading the Bible and copying manuscripts had to have very good vision. Because he spent so much time reading, any vision problem he did have would be quite noticeable, and symptoms of eyestrain and headaches would become evident and bothersome. A warrior or tiller of the soil could accomplish his daily tasks without perfectly acute vision, and the use to which he put his eyes would tend to be less demanding and thus less likely to cause secondary symptoms.

Today, with more widely spread literacy and a greater range of things that it is important to see, many people who are not scholars still want to see well and will choose to wear glasses if they need that help to achieve good vision. But it remains a matter of choice.

The four factors that go into making that choice are: how bad your optical error is, whether or not you are bothered by secondary symptoms, how much demand you put on your eyes, and how well you need to see to feel comfortable in the world around you.

Let us consider these factors carefully, because a misunderstanding of any of them or of their relation to each other can be a source of great confusion. Take, for example, a student who has a minor optical defect. Because he reads a great deal, he makes great demands on his eyes. He will notice his vision problem and will probably experience headaches and eyestrain. If he wears glasses to correct his optical defect, he will be able to see well and he will not be bothered by secondary symptoms. Someone who does less reading but has the same or even a more serious optical error may not be aware of a vision problem and may not experience headaches and eyestrain because his visual demands are less. This does not mean that the student is hurting his eyes through overuse. He is not making his optical error worse, and he is not inflicting any damage on the health of his eyes. His optical error simply interferes with his greater visual needs (seeing very clearly all the time), and the greater demand he puts on his eyes is causing symptoms of strain.

Glasses are essentially just a prosthetic device, like false teeth, which are not essential to your health but nice to have if you want to be able to chew certain foods. Or like elevator shoes for short people. There is nothing unhealthy about being short, but if it bothers you, elevator shoes will make you look taller. As soon as you have removed the shoes, of course, the height you gained is lost until you put the shoes on again.

If you have an optical error, even a great one, but you choose not to wear glasses, you will not be endangering the health of your eyes. The only exception to this rule is the case of young children with particular eye problems that could result in an inability to develop normal sight. Although the choice is a free one for the rest of us, most people will opt for the improved vision and greater comfort glasses can provide.

What Types of Glasses Are There?

Eyeglasses are specially ground lenses that are custom-made to the optics of the individual eye. Their function is always to deliver a focused image to the retina. Eyeglass lenses are worn outside the eye and form a surface through which light passes on its way from the object to the eye. The optical center of the lens is a tiny point positioned directly in front of the pupil. When light passes through the optical center, it is bent in a way that compensates for the eye's inability to bend the light correctly. And that's that. Glasses do not strengthen or weaken the eye, nor do they affect any sort of cure or change that lasts after the glasses are removed.

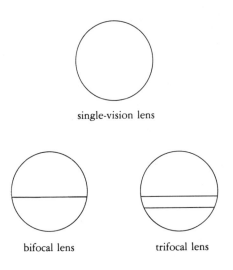

single-vision lens

bifocal lens   trifocal lens

Three basic lens types.

The most commonly prescribed type of eyeglass is called the *single-vision lens.* This does not mean it is for seeing at one distance only; it can be a reading glass, a distance glass, or it can be worn for seeing both close up and at the distance. Single vision simply means that there is only one set of optical corrections in the lens. This distinguishes it from the second most common type, the *bifocal.*

Bifocals contain two sets of optical corrections, one at the top and the other at the bottom of the lens. The most common application for bifocals is for people who are presbyopic and have another optical error that needs correcting for good vision at the distance. The top portion is for viewing distant objects and contains the correction for the distance optical error only. This can be nearsightedness, farsightedness, or astigmatism. The bottom portion is for reading; it contains a combination of the presbyopic correction and the correction for the other optical error, the one represented in the top portion.

Although it usually takes a while to get accustomed to bifocals, most people are able to use them comfortably, since we tend to look down when we read and straight ahead when we look at objects farther away. By keeping the eyeglass centered in front of our eyes and just moving our eyeballs, we are able to look through the proper portion of the lens. The most frequent adjustment problems bifocal wearers have occurs when they walk up or down stairs. The difficulty here is that the distance between eyes and feet is great enough to require distance correction, but we look down (or into the near-vision part of the lens) to see where we are stepping. I advise my bifocal patients to bend their chins down when using stairs or stepping off curbs so they can look through the top of their bifocals. Developing this habit usually solves the problem.

Bifocals are never absolutely necessary, but they are convenient since they make it unnecessary to change glasses every time you want to change from near to distance vision and back. However, some people think bifocals will give away their age, so they prefer having two pairs of glasses and will put up with the inconvenience. The inconvenience can be considerable for

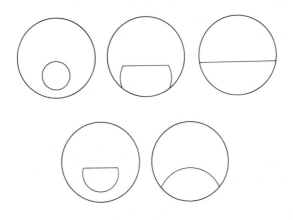

Some types of bifocal lens shapes.

people whose daily activities require that they see well at both distances in close succession.

There are numerous shapes available in bifocal correction. One type has a small rounded portion near the bottom for near correction. In recent years a lens with a relatively large near-correction area has become more common. Both these types have a visible line separating the near and distance areas. Bifocals with no obvious division between the two portions are also available. The demarcation line is ground down so there is a blurred area seen by the wearer but not evident to the outside world.

I decide which type to prescribe, depending on the individual patient. For example, if I have a patient whose visual needs suggest bifocals, but who is reluctant to don this badge of middle age, I can prescribe the last type. However, I will caution this patient that the blur is quite difficult to get used to and that many people find this kind of bifocal not worth the trouble. If appearance is not an issue, I would prescribe one of the types with a clear line of demarcation, and, all things being equal, I prefer the type with the large near portion. A larger close-viewing area clearly separated from the distance portion is, I have found, the easiest bifocal to adjust to. I should say, however, that most of my bifocal patients have no adjustment difficulty at all.

There is another kind of bifocal that can be used by people who are presbyopic but have no other optical error. The bottom contains the correction for their presbyopia, the top is plain glass with no correction at all. The wearer looks down and through the presbyopic correction; when the wearer looks up, he or she looks through unground glass. Half glasses are another version of the same principle. In this case, the wearer looks through air when he or she looks up. In both cases, there is no correction in the upper portion because the wearer has no distance vision problem. The advantage is that the glasses need not be removed every time the wearer wants to see clearly in the distance.

Half glasses for reading.

Trifocals give the wearer correction at a third additional viewing distance—the so-called middle distance. Although trifocals work well once one has become accustomed to them, my experience has been that the adjustment is difficult for many people. I am reluctant to prescribe trifocals unless a special occupational need exists. A bookkeeper, for example, must be able to focus on three different distances constantly and interchangeably: close up to make entries in a ledger; in the middle distance to read figures from papers or to work

with an adding machine farther away on the desk; and in the distance to see across the room. For such a person, the constant need for excellent and comfortable vision at three distances is worth the trouble of adjusting to trifocals. But there are few occupational needs that would justify this inconvenience.

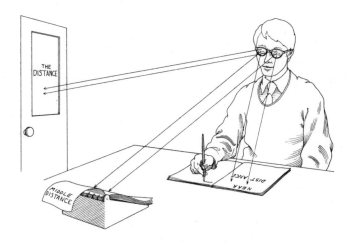

A bookkeeper wearing trifocals.

Most other occupational needs that require middle-distance vision do not require good near and distant vision at the same time. A pianist or other musician is a good example of this. Although it is true that reading music involves middle-distance viewing, musicians do not need to see close up and at the distance at the same time as they read their music. They focus always on the music, rarely, if ever, looking at their hands or at the audience. For such a person, I would prescribe a pair of single-vision lenses with a correction for the middle distance to be worn when playing his or her instrument. This will allow a wide range of good vision within that distance, rather than the narrow middle distance permitted with trifocals.

Graduated focal-length glasses, a variation of the no-line

bifocal, work on the same theory, but they are intended to provide an even greater range of focal distances: far in the distance at the top of the lenses and then gradually closer until the bottom, where correction for near vision is provided. The only trouble is that they do not work very well. Patients have enormous difficulty getting used to these glasses, and I cannot conceive of a single individual patient need that would justify their use and considerable expense.

There is a simpler, less expensive, and less troublesome way to provide a wider range of viewing distances which I often suggest to those of my patients whose presbyopia is still getting worse. When I give them a new prescription for reading glasses, I recommend that they keep their old pair of glasses for viewing middle-distant objects. The rationale behind this is that at the same time as they have come to need a stronger pair of glasses for reading at fourteen inches, they have also lost their ability to focus without correction at, say, twenty or twenty-five inches. Although their old glasses will no longer permit them to focus at fourteen inches, they will make good vision possible at twenty or twenty-five inches. And they may find it useful to wear that old pair of glasses when they play cards, work in the kitchen, or engage in other activities that require them to see well in the middle distance. I am by no means recommending that everyone purchase a separate pair of glasses for middle-distance viewing, but if you already have the glasses, you may as well keep them on hand for these special situations.

Who Needs Glasses?

People who fulfill most or all of the four interrelated conditions: they have an optical error; they find their less-than-perfect vision enough of an annoyance to want to improve it; they put demands on their eyes that make their optical error a hindrance to how well they need to see; and the demand causes them to experience secondary symptoms of headache and eyestrain. Optical error alone, regardless of how small or great, does not mean you have to wear glasses.

Each optical error is corrected with a specially shaped lens, and in each case the result is a focused image on the retina. But let's look at how each specific error is corrected with glasses and when the glasses need to be worn, keeping in mind that I am talking always about the person who has chosen to wear glasses, not simply the person with that particular optical error.

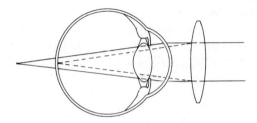

A convex lens pushes the focused rays of light forward to the retina to correct farsightedness.

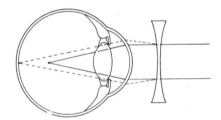

A concave lens pushes the focused rays of light backward to the retina to correct nearsightedness.

If you are nearsighted, eyeglasses will permit you to see well close up and at the distance. The shape of the lens is *concave,* and its effect is to push back the focused rays of light farther than the eye can do itself. This works no matter what the cause of the nearsightedness is, and it works always in the same way.

When and how often nearsighted people wear their glasses depends on how nearsighted they are and on what their visual needs may be. Glasses will always need to be worn for them to see well in the distance. People who are only slightly near-sighted may thus wear glasses for driving (when acute distance vision is not only essential but legally required) and for watching movies and television. They may not need them for closer distances or for distance viewing when acute vision is less important to them. For example, walking around the house or even on the street may not be, for some people, a situation in which seeing with absolute clarity is essential.

Very nearsighted people may not be able to see well beyond the tips of their noses and will thus wear their glasses all the time. Moderately nearsighted people also tend to wear their glasses all or most of the time, particularly if their activities require frequent interchangeable near and far vision. Even if they can read and see well close up without their glasses, they will not be harming their eyes or subjecting themselves to secondary symptoms if they do wear them. In fact, without their glasses they would be utilizing their optical error of nearsightedness, rather than their near-focusing muscles, to focus on near objects. With glasses, their nearsightedness is corrected and they must use their focusing muscles as people with normal uncorrected vision do. In this case, the choice to wear glasses all the time is made because it is more convenient than constantly removing and replacing them.

As a doctor, I would not urge a patient one way or the other. Even though it is true that wearing glasses and using the focusing muscles is more physiologically normal than using an optical error for focusing, it is not medically healthier.

A *convex lens* is used to correct farsightedness. The effect here is to bring the rays of light into focus sooner than they normally would be, so that a focused image is delivered to the retina. And again, the degree of convexity depends on the degree of farsightedness.

If you are farsighted, your decision as to whether or not to wear glasses depends on your age as well as on the four factors that apply to all other optical errors. Take, for example, a young farsighted person who can use his near-focusing muscles to correct distance vision. His focusing power may be great enough and his optical error small enough for him to see well both at the distance and nearby without experiencing any symptoms of headaches or eyestrain. Even though he is farsighted, he will not be motivated to wear glasses, since his vision is fine without them. A person with a greater optical error and/or somewhat reduced focusing ability may experience eyestrain when he reads and therefore will want to wear glasses for reading only. A person whose optical error is so great and/or whose focusing ability is too weak to allow him to see well both at the distance and nearby will experience secondary symptoms all the time and will want to wear glasses all the time.

What about the older person who has no focusing ability at all? If the farsightedness is great enough to cause problems in the distance, this person will have to wear distance glasses, which will correct the optical error of farsightedness, and reading glasses, which will correct both the presbyopic error and the farsighted error. This person is a candidate for bifocals.

The shape of the lens used to correct astigmatism depends on the type of astigmatism you have. If it is farsighted astigmatism, a convex lens is used; if it is nearsighted, a concave lens is used. Astigmatic lenses also have a correction ground into them to offset the error in slope of the cornea on whichever plane it occurs. This correction collapses the two images into one and then the curved lens delivers that focused image to the retina.

Depending on the degree and type of astigmatism, one may need glasses just for reading, just for distance, or all the time. The decision is based on when the blur and discomfort symptoms are appreciable enough to be a nuisance.

The shape of the lens for correcting presbyopia is convex. People who are presbyopic need the correction only when they are viewing close objects, since this condition creates no distance viewing problems.

How to Buy a Pair of Glasses

If your eye doctor prescribes glasses, the next step is up to you. Ophthalmologists do not make glasses; glasses must be ordered through an *optician,* a specially trained and licensed technician who dispenses lenses according to your eye doctor's prescription.

If you do not know where to go for your glasses, you can ask your eye doctor to recommend a particular optician. Or you can follow these guidelines: In general, there are three classes of opticians. Which you choose depends on how complicated your prescription is and how much money you are willing to pay for your glasses.

The best opticians, who also tend to be most expensive, are usually members of the Opticians' Guild. Membership in the Guild is not a guarantee of quality, but it is a good indication that that particular optician is at the top of the field. Guild opticians display their Guild membership in their windows or indicate that they are Guild members in their telephone-book listing. You can also call the chapter of the Opticians' Guild in your state or city to get the names of member opticians in your area (see Useful Addresses on page 209). I always recommend that my patients who have problem prescriptions get their glasses through Guild opticians, although this is less important for those whose prescriptions are not so complicated.

The second type is what I call the neighborhood optician. Although not members of the Guild, these local merchants are usually quite skilled, and they generally charge less than Guild opticians. If a neighborhood optician has a good reputation among people you know, you should feel confident about buying your eyeglasses from him.

The third type, the discount optical house, is the least desirable, and I try to steer my patients away from these outlets. They are often chains; they frequently advertise on television, radio, and in newspapers, offering huge selections of fashion frames and fast, often same-day, service. They depend on volume sales, and their work tends to be sloppy. The frames and lenses they sell are usually not of the same high quality as those available through Guild or neighborhood opticians, and high-pressure selling often goes along with the promises of low price and high fashion. Furthermore, the salesperson who fits glasses at these outlets is most likely not an optician at all.

Purchasing glasses is a serious matter requiring careful consideration and expert, objective advice. Glasses must serve your daily visual needs, and you must feel comfortable with the appearance and fit of both lens and frame. Since glasses are an expensive investment, it is important that you make the purchase calmly, carefully, and without being subjected to pressure. This is often difficult at the discount houses, but I would like to recommend that, wherever you purchase your glasses, you make certain *you* are satisfied—with the glasses themselves and with the price you are paying. Take your time and don't allow yourself to be rushed. Take along a friend if you feel you need a second opinion on how the frames look. This is particularly valuable if you see so badly without your glasses that you cannot tell how you look in a particular pair of frames.

A common experience for shoppers at discount opticians' shops is that they are talked into buying additional pairs of glasses that they do not need. If you require more than one pair of glasses, your eye doctor will discuss this with you and give you a prescription for each. When you go to purchase glasses, you will be armed with any and all prescriptions you need. Do not be badgered into buying additional glasses. The only case in which you might want an additional pair is if you have decided to purchase sunglasses. If you are interested in prescription sunglasses, discuss this with your doctor so he can write a prescription that suits your outdoor viewing needs.

If you find yourself being pressured by an optician, leave the store. The optician should return your prescription to you at your request, but if this is awkward, simply call your eye doctor and ask him to send you another one or have your new optician telephone him for it. Similarly, if you are confused in any way by the contradictory advice of an optician or about any other matter regarding the purchase of glasses, do not hesitate to call your eye doctor before proceeding with your purchase. It is also a good idea to find out what the total cost of your glasses will be before ordering them.

What happens when you get a prescription for a new pair

of glasses and they do not feel comfortable? It usually takes a few days to get used to a new prescription. I advise my patients to let a few days pass before they decide that there is something wrong. If you are still uncomfortable after that, go back to the optician and ask him to check if the prescription was correctly made and if the glasses are properly centered. If everything is as it should be but if they still don't feel right, go back to your eye doctor so he can recheck the prescription.

After you select an optician, there are a number of other decisions to make in your choice of glasses. Corrective lenses can be made up in glass or plastic. I usually recommend plastic to my patients. Most states require that glass lenses be made of shatterproof glass, but plastic lenses are absolutely unbreakable. They are also lighter, which is particularly important if you have very thick glasses or have chosen a large frame. Plastic lenses may add to the cost of your glasses, but I think they are worth the difference.

Although plastic lenses do scratch more easily than glass ones, scratches can be avoided if proper care is taken. Scratches occur only when you allow the lenses to come in contact with a hard surface. If you do not drop them or lay them down on a table or other hard surface, and if you are careful when you clean them, the chances of their becoming scratched are greatly reduced. Plastic lenses will not become scratched from normal wear—they get scratched only when someone or something scratches them.

You will also have the option of buying tinted or clear lenses. Tinted lenses are of no medical benefit. Some people like them for their cosmetic effect or because certain tints cut down somewhat on glare. I have never urged a patient to get tinted lenses, but I would not discourage it either. Tinting adds to the cost of a pair of lenses, and it is up to you to decide if they are worth the additional cost.

Sunglasses are heavily tinted glasses meant to be worn outdoors as protection from the sun's glare. They can be made up with or without optical correction. There are various intensities and various colors available in sunglasses, and which you select is a matter of personal preference. No particular color has any medical superiority over another, nor can any color be harmful to your eyes. It is a good idea to shop for sunglasses on a sunny day; that way, you can test the various intensities and colors outdoors and select the one most comfortable for you.

Sunglasses that become darker or clearer, depending on available light, have become popular in recent years. Their appeal has faded somewhat, however, because they are more expensive and do not work terribly well. The chemical reaction within the lens that causes it to darken or lighten takes a bit of time. What tends to happen is that a person suddenly going into a brightly lighted place has to wait longer than is comfortable for the lenses to get dark. When one goes inside, the lightening action also takes too long. Furthermore, the darkest the lenses get is not very dark, and they are never completely without tint. These glasses are certainly not harmful to the eye, but our irises and retinas do a better and quicker job of adjusting to available light, and they're free!

Sunglasses with graduated tints—dark at the top and gradually lightening until they are almost clear at the bottom—are of no optical or medical benefit. The choice of these glasses is a matter of fashion and taste, but do test them out in sunlight before you buy them, since you may find they do not provide sufficient protection against glare to suit your needs.

The most economical way to get the benefits of prescription sunglasses is to buy a pair of tinted clip-ons. Available in drugstores and variety stores at reasonable prices, these sunshields do not, of course, contain any prescription, but because they can be attached to your regular glasses, they provide protection against bright sunlight while your own glasses provide optical correction.

If you are in the market for nonprescription sunglasses, you need not be afraid of ruining your eyes with a cheap pair from your local drugstore or variety store. A more expensive pair is a better buy simply because sturdier frames will probably last longer and fit better, and a better optical quality lens will give less distortion. Sunglasses are never a matter of medical necessity, but if you feel the personal need for them, by all means invest in a pair.

The frames you choose are also a matter of taste. The most important thing is to select frames that look well on you and will be suitable for your life-style, fashion sense, and comfort. I recommend the sturdiest possible frames, which will last longer and stand up better under constant use. If your preference is for thin wire frames, realize that you make this choice at the expense of sturdiness.

The size of the frames you choose is also a question of personal preference. Oversize frames have become quite fashionable lately, particularly aviator-style frames. There is nothing wrong with wearing large glasses if you like the way you look in them, but they are usually more expensive, and in the outer reaches of the lens the distortion tends to be quite great. Do not expect to be able to look far out to the side or down in the corner of an oversize lens and see a clear, focused image —it is simply too far from the optical center of the lens.

Some Tips on Eyeglass Care and Maintenance

Once you have your glasses, a certain amount of care is required so they will continue to provide you with good visual correction.

The most obvious point is that you should keep your glasses clean. Oil from your skin and dust and grit in the air will cling to your eyeglass lenses. You should clean them off whenever they need it, but certainly at least once a day. The best way to do this is with a clean, soft cloth or one of the special lens-cleaning papers that are available at drugstores and opticians' stores. Some opticians recommend that glasses, especially the plastic ones, be cleaned when they are wet, in order to reduce the chance of scratching. You can use special lens-cleaning fluids for this or plain tap water with or without a bit of soap.

It is also a good idea to keep your glasses in a case when you are not using them, particularly if you are carrying them in a purse or in your pocket. The case will keep the frames from getting bent, protect the glasses if they are dropped, and keep the lenses from getting scratched. If you take off your glasses when you are at home and do not put them in a case, lay them on their sides, resting the frame on the table. Never put them lens side down.

The correct way to put your glasses down.

If your lenses do become scratched enough to interfere with seeing, there is nothing you can do about plastic lenses, but glass lenses can sometimes be repolished and recoated. This works only if the scratches are not too deep, but it is a possible remedy.

Ill-fitting or loose frames that slide down your nose or sit askew on your face will not allow you to get the best benefits of your prescription, since the optical center of the lens must be directly in front of your pupil. This is why a careful and

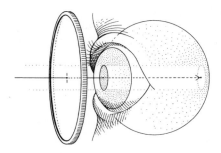

The optical center of your eyeglass lens, the only point of perfect correction, should sit directly in front of the center of your pupil.

conscientious optician will spend what may seem like an excessive amount of time adjusting your frames. If you drop your glasses, or the screws in the hinges become loose, or the frames get bent, you should have them readjusted and repaired so alignment is again perfect. And if you have wire frames, you should have the alignment checked periodically, since this type is particularly prone to bending. Do not continue to wear glasses that are askew, and do not attempt to repair them yourself with paper clips or bits of wire. In most cases, the optician from whom you bought your glasses will make these repairs and adjustments free, but in any case the charge will be nominal and definitely worth it.

I advise those of my patients who are very dependent on their glasses to keep an extra pair on hand. This is obviously a good thing if you lose your glasses and have to wait a few days to get a new pair made. The most economical thing to do is to save your last pair of glasses, which will be adequate for a short time. If you have the money to spend, you can certainly buy two pairs of glasses in your current prescription, but this is not really necessary. It is also not a bad idea to keep a copy of your prescription, especially if you are traveling. The "language" of eyeglass prescriptions is universal; opticians from Paris to Hong Kong can make you a new pair of glasses as quickly and easily as can their counterparts in your home town.

Eyeglasses and physical sports are not terribly compatible. If you are a very active person whose eyes require correction, you may find contact lenses better suited to your needs. But if you prefer to wear glasses, a number of precautions can be taken. There are elastic eyeglass holders that fasten around the back of the head to keep the glasses from falling off. Many people who play tennis, baseball, and other active sports find these helpful. If you swim a lot and find it essential to see well (this is particularly important for underwater divers), you may wear goggles over your glasses. If you are willing to spend the considerable amount of money involved, you can have a face mask made with your eyeglass prescription ground into the glass. Skiiers should certainly wear goggles over their glasses.

Elastic eyeglass holders keep glasses in place during active sports.

Specially ground glass can be fitted into ski goggles, but this, too, is quite expensive.

Camera-viewing lenses and binocular, telescope, and microscope lenses can also be ground to your prescription. But the justification for the expense would have to involve an unusual occupational need. Contact lenses avoid all these problems and are usually a better solution in these cases.

Children Who Wear Glasses

The greatest problems I have experienced with children who wear glasses originates with the parents, not the children themselves. Even though they may not intend it, parents often communicate negative messages about wearing glasses to their children. This is unfortunate, since young children are often quite willing and sometimes even eager to wear glasses. They are quick to understand that glasses will allow them better vision and thus give them an easier time at school and at play. It often makes them feel more grown-up since they see adults wearing glasses. The idea that there is something wrong is usually picked up by children from their parents. I try to encourage parents not to gasp and otherwise express shock or disappointment in their child's presence when and if glasses are indicated. For the sake of the child, parents should try to be positive about the visual benefits glasses will afford. The optical error may decrease or disappear altogether and the child may not need glasses by the time he is a young adult. If the optical error remains, the child will soon be old enough to care for contact lenses and can be fitted for them. Or he may become accustomed to wearing glasses and prefer those in order to enjoy better vision.

If you are a candidate for eyeglasses (and in the case of reading glasses, we all are), I hope this discussion has been helpful in dispelling incorrect notions about what glasses can and cannot do. Although it is true that dependence on eyeglasses can be a nuisance—and often a considerable one—they do make it possible for anyone with an optical error to enjoy normal vision.

Contact lenses are another option open to most people who require visual correction. Many of you may find them a more attractive solution to your visual problems.

4

Contact Lenses

MYTH: Contact lenses are not really safe, and lots of people have trouble wearing them.

FACT: Contact lenses are every bit as safe as eyeglasses, and the overwhelming majority of people who try them can wear them with no difficulty at all. As long as your contacts are fitted by or under the direction of a medical eye doctor, are hygienically cared for and inserted and removed according to your doctor's instructions, and are checked regularly by your eye doctor, there is no reason that you cannot wear them successfully.

MYTH: Contact lenses prevent nearsightedness from progressing.

FACT: Contact lenses, like glasses, can do nothing to alter the internal optics of the eye. All they can do is correct optical errors while they are being worn. This myth has probably arisen because nearsightedness frequently stops getting worse at the end of the growth period and this is the age at which, coincidentally, many people decide to try contact lenses.

MYTH: It is dangerous to wear contact lenses because they can slip off the cornea and float to the back of your eye.

FACT: Because the conjunctiva prevents access to the back of the eye, it is impossible for a contact lens to get there. If a lens does slip off your cornea, it will fall out of your eye or float on your sclera, which will not cause damage.

MYTH: Contact lenses cure astigmatism.

FACT: Astigmatism cannot be cured; however, contact lenses do have an intrinsic correction for the condition. Because astigmatism is caused by an irregularity in the surface of the cornea, it is eliminated while contacts are worn because the perfect curve of the contact lens replaces the astigmatic cornea as the light-bending surface.

The question I am most frequently asked by people who are interested in contact lenses is, "Are they safe?" Many people realize that contact lenses will give them a more natural and attractive appearance than glasses do; others know that contacts can provide them with a better correction of their optical error. But the safety question still arises, probably because people have heard at least one contact lens disaster story and have seen at least one person struggling with a contact lens on the street or athletic field. What they, and perhaps you, do not take into account are the thousands of normal-looking people who pass them each day but who are among the many millions who safely and comfortably enjoy the visual benefits of this technological marvel.

But let me answer the question. Contact lenses are completely safe *if* three cardinal rules are followed: the lenses must be fitted by or under the supervision of a medical contact lens specialist; hygienic insertion and removal procedures must be followed to the letter; and both eyes and lenses must be checked periodically by a medical eye doctor. These rules are easy enough to follow, so there is no reason a person who wants contact lenses and whose optical error lends itself to correction with them cannot safely wear contact lenses.

Contact lenses were first developed in the late 1930s and 1940s, when a glass scleral lens, which completely covered the exposed part of the eyeball, was introduced. These lenses were uncomfortable (five hours was the longest they could be worn), and they were not a significant improvement over glasses. In the 1950s, however, when plastic lenses started to be widely used, the future of contact lenses was assured. The plastics from which they were made were considerably lighter than glass and would not shatter in the eye. Because the lenses were comfortable enough to be worn full time, they became a viable substitute for glasses.

As research and development in the field have continued, contact lenses have become smaller, thinner, and lighter. New materials have been introduced, and the use of contact lenses for a variety of purposes has become widespread.

How Contact Lenses Work

Contact lenses float on a layer of tears in front of the cornea and are held in place by surface tension. Since the lenses do not rest directly on the corneal tissue, they do not interfere with its supply of oxygen. Each time you blink, tears wash under the contact, bringing a new supply of oxygen to the cornea. The tissues of the cornea are strong enough not to be injured by a minute, feather-light lens floating on it, and should the lens slide off the cornea, the sclera and conjunctiva are resistant to damage and abrasion from the lens. In addition, the conjunctiva closes off the back of the eye, so there is no danger that the lens will get lost in your eye and migrate back inside your head.

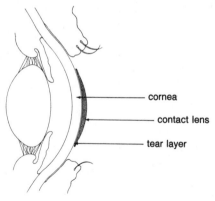

cornea
contact lens
tear layer

Contact lens in position.

I should mention that all contact lenses, regardless of what they are called and how they are advertised, operate on this same principle. Do not be fooled by a particular manufacturer's claims of superiority because a lens is "contactless," "free-floating," or "air-vented." These are simply fancy ways of saying that the lens makes no contact with the eyeball tissue or that it floats on the tear layer or that it allows oxygen to reach the cornea—all of which is true, but certainly not a unique feature. The "contact" in all contact lenses is with the tear layer and not with the cornea, and all lenses permit oxygen exchange through your natural tear fluid. The only significant difference in contacts is whether they are hard or soft, since the plastics from which each type is made, the fitting techniques, and the wearing procedures differ.

This is not to say that it does not matter where you get your lenses and who fits them. That matters very much. The fitting of contact lenses requires very precise measurements with special instruments and an examination of the health of your eyes. This is a medical procedure best performed by a trained eye doctor. A contact lens technician associated with your eye doctor may perform some of the fitting and instruction for care of contacts, and I would not recommend against such an arrangement, but I do strongly urge you to avoid getting lenses from a nonmedical contact lens fitter, who is neither trained nor licensed to perform many of the examination and diagnostic procedures that are essential to complete eye care. The chance that you will get an ill-fitting pair of lenses, which will cause discomfort at best and at worst can damage your eyes, is increased when an unsupervised, nonmedical practitioner does the fitting. The apparent lower cost is a false economy, because the likelihood of a good fit is diminished and the risks to eye health are increased.

If you really want contact lenses but cannot afford a private ophthalmological fitting, I suggest you go to a medical eye clinic at a teaching hospital that offers contact lens fittings. Here, ophthalmologists-in-training work under supervision. In general, the treatment is less personal, waiting times may be longer, and much of the convenience of having a private doctor is sacrificed, but the cost is generally lower than that charged by the contact lens mills and other undesirable non-medical sources, and the quality of medical care is high.

Not all ophthalmologists fit contact lenses, though some work in association with a contact lens technician, who does the actual measurement and fitting of the lenses. If your eye doctor does not fit contacts, you may ask him to refer you to an ophthalmologist who does, or, if you prefer, you may try to find one yourself. The two best sources for this information are the department of ophthalmology at a nearby teaching hospital (at your request, they will be able to give you the names of those ophthalmologists on their staff who have private practices and fit contact lenses) and the Contact Lens Association of Ophthalmologists, a national organization of contact lens specialists. They, too, will be glad to furnish you with two or three names of doctors in your area. (See Useful Addresses on page 209.)

Contact Lens Types

Of the two main types of contact lenses in use today, the most popular are hard plastic lenses. Today's hard lenses are smaller and thinner than they used to be, and they have a higher fitting success rate than contact lenses of a decade ago simply because there is less plastic in the eye.

In the past few years there has been much publicity about soft lenses. Hailed as easier to fit and adjust to, they are made of a kind of plastic that is capable of absorbing water, which gives them the consistency of firm gelatin. They can be flexed between the fingers and are larger than hard lenses. Soft lenses do represent a major advance in the fitting of contacts, and it is a rare person indeed who cannot be fitted for some kind of contact lenses today. But each type has its advantages and disadvantages, and the choice of hard or soft depends on your individual needs.

The main advantage of hard plastic lenses is that as a rule vision is better with them than with soft lenses. Because the soft lens is malleable, a slight bending tends to occur every time you blink. This results in waves in the surface of the soft lens that can create an optical distortion that does not occur with hard lenses, which retain their shape at all times. Hard lenses are also cheaper and easier to take care of than soft lenses.

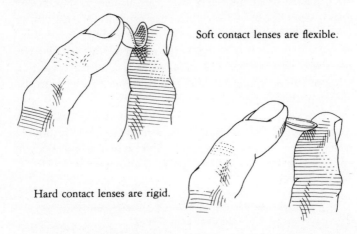

Soft contact lenses are flexible.

Hard contact lenses are rigid.

The great advantage of soft lenses is their relatively immediate comfort. Hard lenses require an adaptation period, during which wearing time is gradually increased until all-day comfort is achieved. Soft lenses can be worn irregularly—once a month, for a few hours at a time, on special occasions only, or whatever wearing schedule suits you. This is impossible with hard lenses, which are intended for full-time—or at least regular—wear. If you wear them less than full time, you must keep them in your eyes more or less the same number of hours each day; any major disruption of that schedule means that you have to readapt to the lenses all over again. Once you have established a regular schedule of lens wear, you must be enthusiastically prepared to wear your hard lenses according to that schedule; you cannot expect to be able to wear glasses on some days, lenses on others, depending on your whim. If you want contact lenses for special occasions only, you should be fitted with soft lenses.

Soft lens wearers rarely have difficulty with foreign bodies in their eyes as hard lens wearers sometimes do. With hard lenses, minute particles of dust or soot occasionally float under the lens and cause irritation to the cornea. In most cases, your eye responds by secreting additional tear fluid, which in a short time washes out the particle. With soft lenses, a larger area of your eye is covered and the flexible edge of the lens tends to adhere more closely to the tear layer, so it is unusual for a particle to wash under the lens and make contact with your cornea. Also, soft lenses virtually never get decentered, but hard lenses sometimes do.

In general, soft lenses provide improvements in comfort, but cost more and require more complicated care. The most important disadvantage, however, is their frequent optical inferiority to hard lenses. Although preferences vary from doctor to doctor, I personally prefer hard lenses, all other things being equal. I think the longer initial period of adjustment required with hard lenses is a small price to pay for the superior optical correction. This comfort factor is really quite negligible, since most wearers adjust quickly to a well-fitted hard lens. Once this period of adjustment is over, hard lenses are every bit as comfortable as soft lenses, and some people find them even more so. Wearers of both types become completely unaware of their lenses—they cannot feel them at all, and only the fact that they see well tells them the lenses are there.

Lens Types of the Future

There are frequent new developments in the contact lens field as attempts are made to improve upon existing materials. Some experimental lenses have been tested and then discarded as impractical or undesirable; others have been more fully developed and may eventually be available to the public. Future possibilities that are still in experimental stages include silicone lenses and lenses made of various new types of plastics. Currently coming on the market are lenses that can be worn full time, twenty-four hours a day. These are a particular boon for older and disabled wearers who cannot easily handle insertion and removal of their lenses. The lenses can be worn even during sleep and are removed for cleaning and checking and then reinserted by the eye doctor during the patient's regular checkup visit.

The newest type of lens available to the general public is the so-called gas- or oxygen-permeable lens. The plastic used in these lenses permits oxygen to pass through to the corneal surface. Theoretically, this augments the oxygen carried to the cornea through the tear fluid. The care, handling, and wearing procedures for gas-permeable lenses are closer to those for hard lenses than for soft. How great an advance over older types of lenses the gas-permeable lens represents remains to be seen.

There are numerous other lens variants being researched, some of which may never reach the market, though others undoubtedly will. The contact lens future is rich with potential.

Glasses versus Contact Lenses

It is popularly but incorrectly believed that the choice of contact lenses over glasses is simply a matter of vanity and, therefore, a costly indulgence. This view is especially unfortunate for men, who could improve their appearance and vision with contacts but are reluctant to do so because they feel it is weak to succumb to vanity. I am glad to see that this sort of thinking is on the wane. I am convinced that the choice of contacts for cosmetic reasons is a positive expression of a healthy ego, and that self-improvement of any sort is to be supported rather than scorned. For that reason I make it a practice to discuss with my patients both the optical and cosmetic advantages that contact lenses have over glasses.

Contact lenses provide better optical correction than glasses in several ways. First of all, they fit in a more natural way, resting directly on the tear layer to create a uniform optical system rather like the natural tear-layer-cornea-lens system we are all born with. With glasses, on the other hand, there is a kind of triple optical system: lens, air, and the optics of the eye itself. In nearsighted individuals, who are the most common wearers of contact lenses, the triple system of glasses makes objects look smaller—considerably so if the person is very myopic—and this can have an effect on vision. An extremely nearsighted person's vision can be corrected to very nearly 20/20 with contact lenses, but objects viewed by that same person through glasses will be made so small that 20/20 vision may not be possible. In practical terms, small equals distant, since the farther away an object is, the smaller it looks and the harder it is to see. Even if the object is made to look smaller by glasses rather than by its distance from the eye, the effect is the same. Contact lenses do not alter the size of objects at all.

Another optical advantage of contacts is that they allow normal peripheral vision, whereas the frames of glasses can obstruct vision on the side. People who wear glasses must move their heads when viewing objects that are not straight ahead to ensure that they are looking through the optical center of their glasses, because if they look through an area toward the edge of the glass they get a prismatic distortion. Contact lens wearers can rotate their eyes and see clearly up, down, and on all sides. The central optical zone of contact lenses is not only proportionately much larger, but it is also constantly in front of the pupil.

Another advantage of contact lenses is in the correction of astigmatism. By creating a perfect front surface of the eye— in effect, by acting as a substitute for the irregularly shaped cornea—contacts automatically correct astigmatism, whereas glasses must have an astigmatic correction as part of their prescription. One of the more frequent reasons for prescription changes in glasses is not an increase in nearsightedness or farsightedness but a change in astigmatism. The need for change, therefore, is reduced when contact lenses are worn because a change in astigmatism does not require a new prescription.

In spite of these real optical advantages the predominant motivation for wearing contact lenses is still cosmetic. It is a fact that people look better and more natural without glasses on their faces no matter how fashionable and attractive eyeglass frames may be. But there is a further cosmetic advantage that you may be unaware of. The alteration of image size works on both sides of the glasses, so the eyes of the wearer, when viewed through their glasses by another person, also look different. This alteration is, in fact, much greater than the one the wearer perceives—a nearsighted person's eyes will appear quite a bit smaller than they really are, and a farsighted person's eyes will appear larger. This is not the case with contact lenses. No image size variation is apparent from either vantage point, so your eyes look normal to the outside world just as the outside world looks normal to you.

Much is made of the discomfort and inconvenience associated with contact lenses, but there is also appreciable inconvenience involved in wearing glasses, though people seem more willing to accept this as a necessary evil. The pressure of eyeglasses on the ridge of the nose, the tops of the ears, and the temples often causes red marks, skin irritation, and headaches. Glasses fog up when you move from one place to another where the temperature is markedly different. Glasses can get wet and misty, just as a car windshield does, but they rarely come equipped with windshield wipers. Contact lenses do not fog up when temperature shifts, and your eyelids serve to wipe your lenses every time you blink. It is true that contact lenses must be cleaned before insertion and upon removal, but most eyeglass wearers find they must wipe their glasses several times a day to keep them clean. I am convinced that contact lenses are, in the final analysis, less of a nuisance than glasses.

The Use of Contact Lenses

In addition to the optical and cosmetic advantages, there are a number of optical and medical conditions for which contact lenses are either the preferred or the only solution. You know by now that corrective lenses of any sort, whether they are glasses or contact lenses, are not in and of themselves healthy to the eyes, since they cannot improve (or, for that matter, interfere with) medical eye health. They do permit better vision if there is an optical error. This application is often called *cosmetic,* though I prefer to call it *nonmedical* to distinguish between purely medical and purely cosmetic reasons for use. Contact lenses can also be used to treat or correct medical eye problems, but most contact lens wearers are fitted for nonmedical reasons, to correct an optical error.

The most common of these is nearsightedness, largely because this condition usually requires full-time correction. Because contact lenses (and particularly hard lenses) are intended for regular, full-time wear, they are ideally suited for nearsighted people. Farsighted people can also wear contact lenses if their farsightedness requires full-time correction, but contacts are obviously not suitable for people who need correction only for reading. If astigmatism, with or without myopia or hyperopia, requires full-time correction, contacts are a good solution.

Patients in the bifocal age group represent a particular correction problem that is solvable with contacts in one of three ways. The least desirable is bifocal contact lenses. These provide optical correction for near and far vision within the same lens. The principle is similar to that of bifocal glasses, but I have found that they do not work terribly well. The rate of successful fittings is quite low, and although there are some practitioners who use bifocal lenses, I have virtually abandoned their use.

The second solution is to fit corrective lenses for distance vision and reading glasses, which can be worn over the lenses, for near vision. The convenience factor is somewhat reduced here, however, since glasses are needed too.

The third solution, the one that I prefer and try out on all my bifocal patients with two healthy eyes, is to fit one eye with a contact lens to correct for distance and the other with a contact lens to correct for near vision. Bizarre and unlikely as it may sound, this setup is remarkably successful and eliminates the need for any glasses whatever. The reason it works is that it allows one eye to see well at the distance but not sharply close up, while the other eye sees well near but not sharply in the distance. Both eyes receive the image, but the one that gets the better image at any given moment tends to dominate, and the brain accepts the clearer message of the two. You are not aware of the blurred image, and unless one eye is covered, you will see perfectly well at all times.

This is a particularly good solution if begun early in the bifocal years, when there is only a slight variation between the correction for near and far vision. As the variation increases, I can make gradual prescription changes, which are not difficult to adjust to.

Purely cosmetic contact lenses are used to change eye color. Larger than standard lenses, cosmetic lenses cover almost the entire cornea in order to obliterate the iris color. The hard plastic lens is opaquely painted to resemble the iris with a clear central pupil through which you see. These lenses are not the same as tinted contacts, which do not change the color of dark eyes but can intensify the color of light eyes. Lens tints are usually chosen not for color change but to reduce glare and make the lens more visible when it is out of your eye. True cosmetic lenses can change any eye to the color desired since the coloring is opaque.

The quality of these lenses varies, along with the price, which is generally high. Inexpensive lenses are quite artificial looking; as the cost increases, the lenses tend to approximate more closely the human iris. Because they are larger than conventional contact lenses, cosmetic lenses are usually less comfortable. People in the performing arts who might, in certain circumstances, require a specific eye color other than their own, are among the few for whom these lenses make sense. But if you have the time, the money, and the whimsical inclination, you can be fitted with cosmetic lenses to match your favorite dress or tie; tiny sequins can be added to the lens to give your eyes an extra sparkle, or if you are in a patriotic mood, you can even get red, white, and blue striped lenses. In general, though, the use of cosmetic lenses is limited by their expense, discomfort, and impracticality. There are a few medical-cosmetic applications for these lenses, however, such as covering a disfigured cornea that has become scarred or opaque. In such cases, the lens is made to match the other eye.

There are several medical and optical conditions for which an eye doctor may prescribe contact lenses as treatment or correction. The most common of these is after cataract surgery. Because removing the cataractous lens creates extreme farsightedness, the image size variation caused by glasses is very disorienting, impossibly so if only one eye has been operated on. Most people find adjustment to contact lenses easier than relearning spatial relations, and hard or soft lenses provide excellent vision without making the eyes look huge and bulging, as they characteristically do with the very thick glasses that are needed after cataract surgery.

Another special use for contacts occurs when there is an appreciable difference between the optical errors of the two eyes. This is difficult to correct with glasses since there is an intolerable discrepancy in the size of images seen with each eye. When hard or soft contacts are used, the problem is eliminated, since both eyes can be corrected without affecting the size of images seen.

There is an unusual, irregular type of astigmatism that cannot be adequately corrected with glasses. In this case, soft lenses are not suitable because the waviness of the cornea causes the lens to bend so much that serious and constant distortion results. A hard contact lens, however, creates a perfect and unchanging surface in place of the cornea and permits the restoration of normal vision when the lenses are in place.

A medical indication for hard lenses is *keratoconus,* a disease characterized by a thinning and forward bulging of the cornea. If left untreated, the disease will progress and result in a tear of the corneal tissue and total scarring of the cornea, leaving the eye with no useful vision. A hard contact lens pressing on the front of the cornea retards the disease process and permits far better vision than would be possible with glasses. The lens functions both as optical correction (since high astigmatism is caused by the bulging cornea) and as a sort of truss to control the bulge.

Soft contact lenses are sometimes used as bandages for certain painful conditions of the cornea. They relieve the pain caused by exposure to the air and protect the diseased cornea from irritation when the eyelid opens and closes over it. This is a purely medical indication and is used only for severe, chronic disease states; your doctor will not use a soft lens bandage for a minor corneal burn or abrasion, for example. No optical correction need be included in the bandage lens, though it can be if necessary.

Optical and medical needs aside, one of the most important considerations in the decision to try contact lenses is your attitude and your sense of determination. As comfortable and effective as lenses can be, success cannot be assured without a firm commitment on your part. If you select hard lenses, you must be enthusiastically prepared to wear them regularly. This does not mean you have to wear them every waking hour (though that is the best plan), but it does mean you must establish a schedule and stick to it seven days a week. Furthermore, the care and maintenance of your lenses is vital to the health of your eyes, and you must faithfully follow your doctor's instruction on these matters. And finally, you should understand that there is a certain amount of adjustment required before you can be comfortable with a foreign object in your eyes.

Because much of this adjustment is psychological, I try to be sensitive to my patients' motivations for getting contacts and to base my decision whether or not to prescribe them to a large extent on that motivation. After explaining to patients who express interest in contact lenses the pros and cons in general and any that are specific to them, I let them decide whether or not they want to be fitted. I never force contact lenses upon an unwilling patient, or even upon one who wavers. During many years of practice I have observed that patient enthusiasm is a major factor in a successful fit. A patient who has no structural or medical problems but who is not fully motivated is much more likely to fail to become adjusted to contacts than a patient who may be more difficult to fit but is determined to wear them.

The Cost of Contact Lenses

Contact lenses are more expensive than glasses, soft lenses somewhat more so than hard lenses because laboratory costs are higher. In general, the cost to you includes your lenses and enough fitting and examination visits to insure a comfortable fit. Many of the commercial contact lens outlets advertise a guarantee of half your money back if you are not satisfied with your lenses; eye doctors rarely offer such an enticement. Why is this? Doesn't this drive people to become customers of these undesirable sources?

The primary reason that eye doctors do not offer guarantees has to do with the issue of psychological motivation. There is no question that spending money represents a commitment to most of us—if you have paid a significant sum for contacts, your determination to succeed is likely to be stronger than if you know in advance that there is a way out, that you can get some of your money back if things are not absolutely terrific. Because there are time limits on a money-back guarantee, you may become impatient and not allow yourself enough time to grow accustomed to your lenses for fear that the guarantee will expire before you know whether or not lenses are for you. You will probably overemphasize minor discomforts and focus on the negative side—the non-fit, money-back idea—rather than on the comfortable fit and improved vision that is most often the case.

But perhaps the most important reason to get your lenses through a medical eye doctor is that he is the only one who is also trained and licensed to spot and treat any medical eye problems that might arise from wearing lenses or that might exist prior to fitting and mitigate against their use. Contact lens stores are in the business of selling contact lenses; eye doctors are in the business of giving complete medical eye care.

The fitting of hard and soft lenses differs significantly, as do the insertion and removal procedures. I begin the hard lens fitting with a complete eye examination, including a determination of the optical correction. Because hard lenses must conform to the size and shape of your cornea (this is what insures a comfortable fit), I also measure the shape of the cornea with a special instrument, called a *keratometer*.

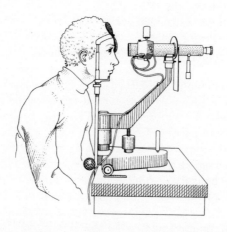

Keratometer—used to measure shape of the cornea.

Once these measurements have been taken, a pair of lenses is ordered from the laboratory, and you are asked to return for a second visit when the lenses are ready. At that time, the lenses are placed in your eyes and checked for proper fit and optics. If everything is satisfactory, you will be given detailed instructions in how to care for and properly insert and remove your lenses. You will be put on a wearing schedule so you can gradually increase the time you wear your lenses each day until you are comfortable with them all day.

The adjustment schedule may vary from doctor to doctor. I recommend that my patients begin by wearing their lenses for two hours the first day and then increase their wearing time by two hours each day thereafter, though it can be less than that if the lenses bother them. It is important that wearing time not be decreased during this period. I see my patients once or twice a week during their initial adaptation period to keep track of their progress, to determine if there are any problems, and to answer any questions they may have.

Once you have become comfortable with your lenses for a full waking day, you should wear them regularly, with as little time variation as possible. That is, if you customarily wear them for sixteen hours daily, do not wear them for more than eighteen or much less than fourteen. If you go for several days without wearing them at all, you must not return to full-time wear without allowing for readjustment. As a general rule, if you have not worn your hard lenses for one full day, you can wear them full time the next day, but if you have been without them for two days, you should begin by wearing them for half your normal wearing time, and then full time the following day. If you have not worn your lenses for four days, begin by wearing them one-quarter of your normal wearing time, and increase by one-quarter each day. According to this formula, it will take as many days to readjust as the number of days you were without lenses.

The fitting for soft lenses begins with the same office visit and the same complete eye examination. The prescription for soft lenses, however, is less complicated than for hard; although the optics of soft lenses vary as much, their softness and flexibility allows them to conform to the shape of the cornea. There are, in fact, only a handful of prescription variations— so few that your doctor can have a stock supply on hand.

On the basis of the optics and corneal measurement, your doctor will select the lens he thinks is most likely to fit you. He will examine the lenses for fit and optical correction, and if his findings are not satisfactory, he will try another pair until the perfect fit and optics are achieved. Once I am satisfied that my patients have the right pair of lenses and know how to insert, remove, and care for them, I let them walk out of my office wearing their new soft lenses.

Whichever type of lenses you have, you should have your

eyes and lenses examined on a regular basis. Your doctor will tell you when to come in for your next checkup; I generally follow a schedule of once every four to six months for lens checks, and once every year and a half for a complete eye examination. These checkups are particularly important because they allow your doctor to keep watch on the medical health of your eyes, and especially your cornea. They are also of value because your lenses can become scratched or chipped in a way that could cause irritation to your eye or injury to your cornea even though you may notice nothing.

The commercial contact lens outfits are usually remiss in this area. Although some do recommend periodic return visits to have the lenses checked, follow-up examinations are by no means a universal procedure. Even when they are urged, an unsupervised lens technician or optometrist is limited in what he can do. Examination and evaluation of the medical health of your eyes is not part of the procedure; a nonmedical technician will, therefore, be unlikely to spot and unable to treat any medical eye problems that might arise.

Using Your Hard Contact Lenses

You have heard me say that contact lenses are completely safe as long as three basic rules are followed. You satisfy one when you get your lenses through a medical contact lens specialist and another when you return for periodic checkups. The third rule is that you follow to the letter your doctor's instructions for proper care and hygienic insertion and removal of the lenses. Both hard and soft lenses have rigid requirements for maintenance, but the specifics differ. In both cases, once you have become accustomed to the drill, it is a marginal inconvenience when you consider the great advantages contacts have over glasses.

I instruct my hard lens patients to use the following technique:

Before beginning, be sure that your hands and the lenses are clean and that you are working in an environment free of the dangers of contamination or loss. Do not try to insert or remove your lenses over a sink. No matter how careful you are to close the drain, a lens can still escape that way, and a lens dropped on the sink or any other hard surface can become scratched or warped.

Spread a clean towel over a flat surface. Place your cleaning solution, wetting solution, a cup of lukewarm water, a standing mirror, and your contact lenses in their case on the towel. You will have to clean your lenses thoroughly before and after each wearing. This is not an over-cautious measure you can ignore or even neglect from time to time. Your lenses will pick up undesirable particles from the air and other contaminants while you are wearing them, and the lenses should not be returned to their case until they have been properly cleaned. In time, these substances can harden on the lens surface, making cleaning more difficult. The sooner the lenses are cleaned the easier it is to do a good job of it. And, although it may seem that the lenses should emerge from their case as clean as they were when you put them there, your case itself is not a perfectly sterile environment and the lenses may pick up contaminants and irritants even there.

The safest way to clean your lenses is to begin by washing your hands twice with soap and water to remove any irritating substances, such as nicotine or body oil, which can cause discomfort. Dry your hands on a lint-free towel and then open your case.

Many lenses have a small dot printed on the right lens to help you tell them apart. You can see the dot when the lens is out of your eye, but you will not notice it once the lens is in place. Whether or not your lenses are marked in this way, it is a good idea always to start with the right lens; this makes it easier for you to remember which is which. Hold the lens gently by the edges and apply a drop of cleaning solution to both sides. Rub the lens lightly between your thumb and index finger to distribute the solution over the entire surface. Avoid using your fingernails, and do not apply pressure to the lens. Dip the lens in the cup of water to rinse off the cleaning solution but do not attempt to dry it. Again holding the lens gently by the edges, apply a drop of wetting solution to each side. You are now ready to insert it.

Place the lens concave side up on the tip of your moistened index finger, and with the middle finger of the same hand pull down the lower eyelid. Raise the other arm over your head and lift the upper eyelid with the middle or index finger—whichever is more comfortable and secure—by grasping the lid near the lashline. Look straight at the center of your lens (it is now in front of your eye and slightly below it). Keeping your finger on the eyelid, move the lens toward your eye until it makes contact with your cornea. At this point release your lid. Close your left eye to make sure that the vision in your right eye is good and that the lens is properly located. Repeat the same procedure for your left eye, using the opposite hands, or if it is more comfortable for you, the same hands. The key to easy insertion is to stay relaxed, keep your eyes wide open, and stare straight ahead.

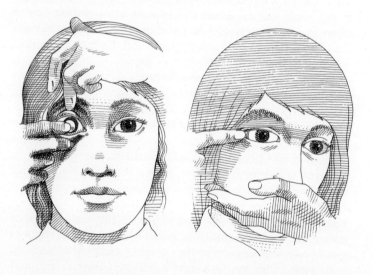

Techniques for hard contact lens insertion and removal.

The removal technique begins with the same setup, whether you are taking out your lenses at the end of the day or because you have a particle or irritant in your eye. There is no need to panic if you have something in your eye, and a hasty and careless emergency removal is one of the surest ways to lose a contact lens.

Once you have set up your towel and lens paraphernalia and have washed your hands thoroughly, lean over the towel and begin by removing the right lens. Place your index finger on the outer edge of your right eye as close to the eyeball as possible and pull the lid to the side while forcing your eyelids to stay open. Cup your other hand under your eye and then blink. The lens should fall out into your cupped hand. If it missed it will land on the towel, which is there as a soft backup and will protect your lens from getting scratched. Sometimes the lens will adhere to your eyelashes, so before you panic and start searching the floor, check to see if the lens is on your lashes. Once you have located the lens, hold it gently by the edges and repeat the cleaning procedure outlined above. Then return the lens to its case and close the case carefully, making sure the lens is safely tucked in and the lid does not close down on the edge of the lens. Repeat the same steps for the left eye.

I do not recommend using soaking solution in the case. Although soft lenses do require special soaking solution to maintain their proper texture, hard lenses can be stored dry without any risk. I feel that soaking solution, aside from being a waste of money, provides an environment for the breeding of bacteria.

Another bacterial danger arises from the foolish habit of using saliva as a wetting solution. Never do this. The mouth is the most bacteria-laden part of the body, and although it is true that your eyes have certain natural bacteria, they are not the same as those found in your mouth. Spitting on a contact lens or wetting it with your tongue risks serious eye infection. If you ever find yourself without a bottled wetting solution designed for your type of lenses, use plain tap water. Wetting solution increases comfort temporarily, but it is quickly washed away by your tears. Water is an acceptable substitute; saliva is most emphatically not.

The cleaning solutions intended for hard lens use will remove small particles and various irritants, but they do not dissolve oils very well. You may, therefore, occasionally want to give your lenses a special cleaning to remove any grease that may have built up. A nonirritating shampoo such as those intended for babies will do this effectively, as will cigarette lighter fluid, benzine, or toothpaste. I know it sounds odd, but these household items work extremely well, and there is no need for you to invest in special "heavy-duty" lens cleaning products. Use these substances just as you would your regular cleaning solution, but do take extra care to rinse them off well before putting the lenses in your eyes.

Within a short time, handling your lenses will become an easy, routine matter. It should not, however, become a matter of carelessness. Strict adherence to these measures will insure that your lenses stay in good condition and that you will be able to wear them with comfort.

It is not necessary, as some manufacturers would have you believe, to carry around a steamer trunk filled with bottles of contact lens solutions. The only product that is absolutely essential is cleaning solution; wetting solution, though certainly helpful, can be replaced with plain water. Totally unnecessary are soaking solutions and the all-in-one solutions that claim to clean, soak, *and* wet your lenses—the fact is, they do none of these very well.

Using Your Soft Contact Lenses

The hygiene and insertion and removal procedures for soft lenses differ considerably from those for hard lenses. The soft lens routine may seem more complicated, but here, too, you can quickly adjust to it and easily integrate it into your day. The specific details vary, depending on the manufacturer of the soft lens in question, so I leave it to your doctor to tell you how best to use the lenses prescribed for you. I can, however, talk about some things that are generally true for all soft lenses, regardless of type.

The absorptive plastic that makes soft lenses flexible must not be allowed to dry out or to absorb anything other than the special solutions intended for use with soft lenses. You cannot use hard lens solutions or any other emergency substitutes.

Soft lenses are stored wet, and, because they could pick up bacteria and other irritants, special cleaning procedures are required. You must also be particularly careful about washing your hands before handling your soft lenses, and if you use aerosol products, I advise that you remove your lenses first. If this is not feasible, be sure to keep your eyes closed while you are spraying and for a minute or two after until no more spray remains in the air.

When you are ready to put in your lenses, wash your hands well and set up a towel, mirror, your lens solution bottle, and lens case on a table. Carefully remove the right lens from the case and examine it to make sure it is moist and clean. Because they are flexible, soft lenses can get turned inside out. Check to be sure this has not happened, and if it has, reverse the lens.

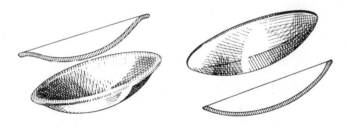

The soft contact lens on the left is inside out; the one on the right is correct.

Clean the lens with the solution your doctor has suggested. This and only this can be used; you may not, as you may with hard lenses, use soap or lighter fluid since these would be absorbed into the lens itself. Place the lens in the palm of your hand with the concave side up. Wet it with solution and rub the lens gently between your thumb and index fingertips. You are now ready to insert the lens.

Place the lens on the outer edge of your index finger, using whichever hand is most comfortable. Hold your head erect and gaze upward at the same time as you pull down your lower lid with the middle finger of the same hand. Keep your eyes fixed on a point above you, but do not move your head up—

the idea is to expose a larger portion of your sclera below the cornea. Gently move the lens onto the sclera, then remove your index finger and release the lower lid. Close your eyes for a moment and the lens will center itself.

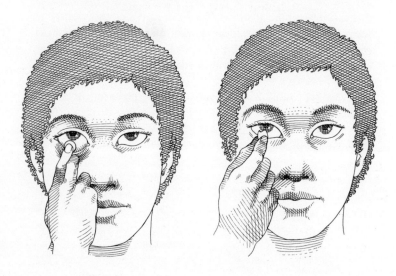

Techniques for soft contact lens insertion and removal.

When you are ready to remove the lens, wash your hands thoroughly and make sure they are well rinsed. Hold your head erect and gaze upward. Pull down the lower lid with your middle finger and put your index finger on the bottom edge of the lens. Gently, without exerting much pressure, slide the lens down onto the sclera. Once the lens is on the sclera, keep your index finger on the lens and continue gazing upward and holding down the lower lid. Place your thumb and index finger on either side of the lens and gently roll them together, pinching the lens so it doubles up between your fingers. This allows air to get under the lens, thus breaking the suction on your tear layer. Once you have the lens pinched between your fingers, remove it from your eye. Clean the lens again, return it to its case, and add some solution.

You can insert and remove your soft lenses in one-tenth the

time it takes to explain the procedure. If you take care to keep your hands free of irritants and always work with lenses that are clean and of the right consistency, you can perform the technique deftly and comfortably.

Special Contact Lens Problems

In the past twenty-five years the sight of a panicked individual clapping hand to eye and running off to the nearest rest room has been a common one. Unfortunately, this has led many people to believe that contact lenses are fraught with danger and distress. But, in fact, if proper precautions are taken, contact lenses can be safely and comfortably worn. Nevertheless, let us look at how best to deal with some of the problems and special situations that can arise when you wear contact lenses.

Hard lenses are vulnerable to scratching and chipping. You may not notice minute flaws in your lenses, but they can cause irritation to your eyes and obviously should be avoided. During your periodic checkups your doctor will examine your lenses to determine if they are flawed in any way, but only you can prevent them from becoming damaged. Avoid all hard surfaces: do not touch your lenses with your fingernails, and if you wear rings, I think it is better to remove them when you handle your contacts; don't bounce your lenses around on mirror surfaces; and take great care not to drop them. This is the main reason that I advise insertion and removal over a towel, not over the sink, for, if dropped, the lenses may become scratched, chipped, or warped when they hit the hard porcelain surface. You cannot scratch your lenses through normal wear; damage occurs only when you do something to scratch them.

If your hard lenses have become scratched, they can, in some cases, be repolished, rebuffed, or resurfaced. If the scratches are severe or if there are chips, the lenses will have to be replaced. But this is not something that will inevitably occur in the course of time—some patients of mine have worn their lenses faithfully for years and, because they have taken proper precautions, the lens surfaces are as smooth as they were the day I first fitted them.

If you do lose a lens, call your doctor so he can immediately order a replacement. It usually takes only a few days, so in the case of hard lenses the readjustment period need not be a long one. Even if you are traveling abroad, you can cable your doctor and he will be able to mail your lenses within a few days. If you can afford it, the ideal situation is to have a spare pair so you will not have to interrupt your wearing schedule at all, but this is by no means a necessity.

In the early days of contact lenses, people often took out insurance against loss of their lenses; this is rarely done today. It made sense at a time when replacement lenses were considerably more expensive than they are now, but current replacement costs are generally low enough so that you would have to lose at least two lenses a year to make insurance premiums pay off. I believe that careful handling of contacts is the best insurance against both loss and damage.

The two most common situations in which lenses are lost need never occur: when a particle of dust enters the eye, causing momentary irritation, and when the lens slips off the cornea onto the sclera. Do not panic in either case, and do not try to remove your lens on Main Street or in the middle of a dance floor. A particle in your eye will usually be washed out with the tears triggered by the irritation. If it is not, calmly and slowly make your way to a quiet corner where you can set up for removal. Do not stand over a rest-room sink. Instead lay a paper towel or handkerchief on a flat surface near a mirror, have your case and lens solution at hand, and proceed in the normal fashion.

If the lens has slipped onto your sclera, it can remain there without doing any harm. You will not see well through that eye since you are no longer looking through the lens, but you cannot damage your eye by leaving the lens on the sclera for a while. Soft lenses rarely get decentered, but hard lenses sometimes do. If this happens, you can rectify the situation in one of three ways:

The first but least effective way is to look in a mirror, locate the lens, and then look in the direction of the lens. By bringing your cornea to the lens, you are trying to get the lens to recenter itself once the particular contours on which it was designed to float are again underneath it. If this does not work, close your eyes and gently massage your eyelid until you have brought the lens back to your cornea. Do not press hard; exert only as much pressure as needed to move the lens gradually toward your cornea. The third method uses your index finger on your eyelid margin. Look in a mirror and move your eyelid

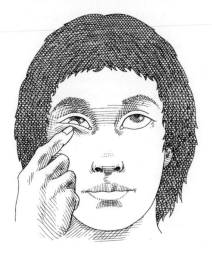

Recentering a hard lens, using index finger and eyelid margin.

(bottom if the lens is below the cornea; top if it is above) so you gently push the lens along until it regains its position on the cornea. Do not, in any circumstances, place your finger directly on your eyeball or on the lens. Many times the lens will drop out of your eye in the midst of these procedures. When that happens, simply clean the lens, reapply wetting solution, and reinsert it. But be sure you are working on a flat surface covered with a towel or soft cloth so your lens does not get scratched if it does pop out.

Some contact lens wearers are disturbed by low levels of humidity, such as those that occur in overheated houses during the winter. The best solution is to add humidity to the air by opening a window, using a room humidifier, or turning down the heat. If this is not practical, moisturizing eye drops, the kind called "artificial tears," can be used.

Glare can sometimes be a problem for contact lens wearers. Lightly tinted lenses reduce glare somewhat, but if you are still bothered, wear a pair of nonprescription sunglasses when you are outside. Sunglasses are also helpful on windy days, especially during the initial adjustment period, because they provide a shield against dust and airborne particles.

Contact lenses, hard or soft, are designed to be worn throughout the waking day. You can engage in any and all activities, regardless of how strenuous. In fact, lenses lend themselves to wear during contact sports, dancing, and other such activities better than glasses do. I do not, however, advise wearing lenses when you swim or bathe unless you wear goggles or are careful to keep your eyes closed or closely squinted. I have had many patients who have lost their lenses this way —a contact lens that floats out of your eye and into a swimming pool is gone forever.

Finally, most contact lenses are not intended to be worn during sleep. You must remove them before going to bed, even when you intend to catch no more than forty winks. Blinking is an important part of what makes contacts safe, and you do not blink when you are asleep.

A pair of well-fitting contact lenses, worn regularly and kept in good condition, will provide the best possible optical correction. Unlike glasses, which create a sort of triple optical system that can distort the size of viewed objects and can, therefore, sometimes make 20/20 vision impossible, contact lenses closely approximate nature's optical system. Their medical and nonmedical applications have made it possible for millions of people to see well and to maintain the health of their eyes. When fitted by or under the supervision of a medical eye doctor and cared for properly, they are a significant improvement over glasses, since they provide you with a far more natural window on the world.

5

External Eye Problems

MYTH: Sties are contagious, but they can be cured by placing a piece of metal, especially silver (a ring or a dime, for example) on your eyelid.

FACT: Sties are not contagious; they do not spread from one person to the next. They are an infection, and a piece of metal will, at best, do nothing for a sty and, at worst, can cause further infection because of germs it may carry.

MYTH: If you have measles, you should stay in a darkened room or you will injure your eyes.

FACT: Measles is a viral infection that can infect the cornea as well as the rest of the body. A corneal infection will make you more sensitive to light, so staying in a darkened room will be more comfortable. But lights will do no injury to an infected eye. As with all illnesses, anything that can be done to make the patient more comfortable is a good idea, but comfort is the only factor.

MYTH: Sties can come from a cold in your eye, so they are a kind of cold sore.

FACT: A sty is not a cold sore, and it is never caused by a virus as colds are. *Cold sore* is the common name for a specific viral infection called *herpes simplex.* It is not the same thing as a sty, nor can you catch a cold in your eye. You can get a viral infection of various parts of your external eye, but this is not the same as a cold, which is an upper respiratory infection.

MYTH: People who care about the health of their eyes shouldn't use eye make-up.
FACT: Properly and carefully applied, eye make-up will not endanger the health of your eyes. If you like the way you look with eye make-up, by all means use it.

―――――――――――――

If our eyes are our window on the world, they are also a doorway through which infection and foreign bodies can pass. Fortunately, the eye's protective system—the eyelids and eyelashes, tears, the conjunctiva, sclera, and cornea—does an excellent job of guarding the delicate interior of the eye from infection and other intrusions. Infections inside the eye, when they do occur, usually come from an infection in the body, and these tend to be serious. But as much protection as the outer layers provide, they are themselves vulnerable.

The common external eye problems are generally not serious. Some of them require no treatment at all and go away by themselves; many others are easy for your eye doctor to treat with special medication. In the pages that follow, I will describe the most common problems and tell you about their causes and symptoms and how they can be treated, but if you find yourself with any of these symptoms, do not assume that you can make your own diagnosis and then run to your local pharmacy for over-the-counter medication. See your eye doctor so he can properly diagnose and treat the condition to ensure that it does not become a serious eye problem.

―――――――――――――

Pink Eye

Pink eye is the common name for *conjunctivitis.* It is used to describe any inflammation of the conjunctiva, the transparent

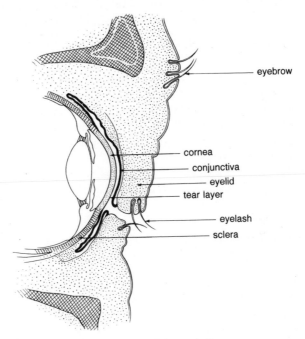

The protective system of the eyeball.

membrane that covers the front surface of the eyeball and laps over onto the inner eyelids. The three main types of conjunctivitis, each named for the cause of the inflammation, are infectious, allergic, and chemical.

Infectious conjunctivitis can be caused by bacteria or by a virus. In both cases, the symptoms are red eyes, inflammation of the inner eyelids, excessive tearing, and a sandy or scratchy feeling in the eyes. People with pink eye also commonly have a discharge from their eyes which makes the eyelids sticky in the morning since the discharge is not blinked away during sleep. If the cause of the pink eye is a bacterial infection, the discharge will be more pus-like; if a virus is responsible, the discharge will be more watery.

I treat bacterial conjunctivitis with antibiotic drops and/or ointment, but viral conjunctivitis cannot be cured with antibiotics. The body's own defense system will work against the virus and in time it will go away. All the same, I often prescribe antibiotics to prevent secondary bacterial infection, since the conjunctiva is more susceptible to bacteria in the presence of a virus.

Infectious pink eye is contagious, so you should take a few common-sense measures to prevent its spread: wash your hands before and after applying eye medication; do not share your eye drops with someone else; keep your hands away from your eyes; and be careful not to share towels or washcloths with others in your household.

Pink eye can also result from an allergic reaction, and in this case it is not contagious. Anything that can cause an allergic reaction elsewhere in your body—such as pollen, cosmetics, cats, dogs, or fabrics—can cause *allergic conjunctivitis.* The symptoms are itching and burning eyes, pronounced redness, and excessive tearing.

The ideal treatment is to remove the cause of the allergy. If this is impossible—either because you cannot identify the cause or because you cannot bear to get rid of your cat—your eye doctor can prescribe one of a variety of eye drops to help alleviate the symptoms.

Some drops contain a mild local anesthetic, a drug to constrict the blood vessels in the eye and thus to reduce redness, and a soothing liquid. Other kinds of drops contain cortisone. There are also a number of over-the-counter eye drops that are advertised as effective against allergies, but I have found these to be weak and of little use. The medication they offer is quite mild, and they often contain other substances that, though they will not damage your eyes, are not at all effective against allergic conjunctivitis.

Unfortunately, medical science has not been very successful in finding a cure for allergies, so the most widely accepted approach is to try to relieve the discomfort they cause. This is particularly important when the allergy affects your conjunctiva because rubbing your itchy, irritated eyes will only aggravate the condition. If you are bothered by allergic eye irritation, see your doctor, not your druggist.

Chemical or *toxic conjunctivitis* is caused by irritants in the external environment. These include high levels of air pollution, noxious fumes, and chlorine in swimming pools. As with allergies, the ideal treatment is removal of the irritant. If this is not feasible, lubricating eye drops to soothe the eyes can be tried. If, for example, you work in a factory where you are constantly exposed to irritating fumes, or if you live in an area with a great deal of smog, nonprescription eye drops—the kind referred to as artificial tears—may be helpful. If highly chlorinated water irritates your eyes, try wearing goggles when you swim.

Sties

The common sty is simply an infection of one of the small glands that lie along the eyelid margin. It is similar to the skin infection we know as acne. Sties first appear as painful red lumps on the edge of the eyelid, and in their later stages a head of whitish pus will appear. If sties occur often, an unnoticed chronic infection of the eyelid may be the cause.

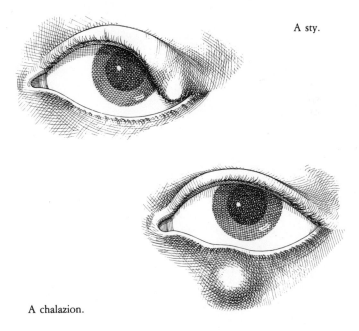

A sty.

A chalazion.

Sties are treated with hot compresses applied to the affected area. A clean washcloth dampened with hot tap water is as effective as anything else, though you may use a boric acid solution if you like. In addition, I often prescribe antibiotic drops or ointment to control infection. If this treatment is begun early, it will in most cases clear up the sty in a couple of days. Only rarely will surgical incision and drainage of the sty be required, and when it is, this can be done in your eye doctor's office.

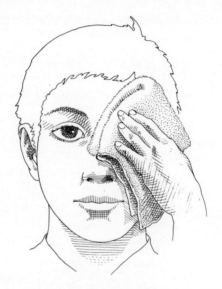

Applying a hot compress.

Sties occur more frequently among school children than among adults. This is not, as you might assume, because children tend to "pick up" infections in school. Sties are not contagious and cannot be passed from person to person. Rather, it is more likely because children's glandular secretions are more erratic, particularly during puberty, and because children tend to be less careful than adults about keeping their hands clean and away from their eyes.

Corneal Infections

The cornea, that clear tissue layer that lies in front of the iris, can become infected in several ways. Its surface can be broken by some direct injury: a foreign object that has become lodged in the cornea; a scratch from a paper edge, a branch, or the sharp edge of a leaf; or an abrasion caused by a fingernail, cosmetic brush, or improper use of contact lenses. And a secondary infection can result. Another common cause of corneal infections is a complication of pink eye. Since the cornea is surrounded by the conjunctiva, an untreated infection of the conjunctiva can easily spread to the cornea.

Normally our tears provide protection against corneal infection. In the course of the day our eyes are constantly bathed with tear fluid and our eyelids act as windshield wipers every time we blink. But some people—and all of us to a degree when we get older—secrete less tear fluid. When this happens, the cornea becomes somewhat more susceptible to infection.

The symptoms of corneal infection are pain, which is sometimes severe; a teary, reddened eye that is sensitive to light;

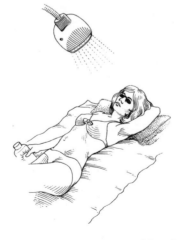

Special protective goggles must be worn while using a sunlamp.

and a scratchy feeling when you blink. Corneal infections are usually not contagious.

The infection can be caused by bacteria or a virus. I use antibiotic drops and/or ointment to treat bacterial infections, and for some viruses specific antiviral drops are available. I will often also patch the infected eye, since keeping the eye closed in this way avoids further irritation and discomfort caused by the lid moving over the cornea with every blink.

A corneal infection can also be a secondary result of a corneal burn, often caused by an ultraviolet sunlamp. The burn itself is extremely painful and needs to be treated medically. I prescribe anesthetic drops to relieve the pain and antibiotic drops to guard against secondary infection. And, again, I will usually patch the affected eye to give it a rest. If both corneas are burned, I will try to patch only one eye so the temporary visual impairment is not total, but in some cases both eyes must be patched.

Scleritis

The white of the eye can become inflamed, usually in response to infections or irritations in nearby tissues. Less frequently, the inflammation originates with the sclera itself. When it does, this is called *scleritis,* and it is most often the result of an allergic reaction and, therefore, not contagious. A much rarer scleritis can be caused by an infection in the white of the eye. The symptoms of true scleritis are a red sclera and slight discomfort; it is usually treated with eye drops that contain cortisone.

Blepharitis

Blepharitis is an infection of the glands of the eyelid margins that bears a similarity to dandruff, and it often occurs in people who have dandruff as well. It is usually a chronic condition, which means it tends to recur and is often present in a very mild form with minor, if any, symptoms. It also appears in an acute form, at which time the infection is active and the symptoms are quite obvious. People with acute blepharitis have reddened and encrusted eyelid margins. The treatment for acute blepharitis is an eye drop that contains cortisone and an antibiotic. Your doctor will probably also prescribe an effective antidandruff shampoo for your scalp and eyebrows. I advise patients who have chronic blepharitis and suffer recurrent attacks to use the shampoo preventively, once a week or so, even when the condition is inactive.

Blepharitis is not contagious, and it is usually mild and easy to treat. If it is severe or neglected, however, abcesses of the eyelids, secondary infections of other parts of the eye, and loss of the eyelashes may result.

Chalazions

A *chalazion* is a cyst inside the eyelid caused by an infection. The infected cyst is painful and reddish; when the infection goes away, the cyst remains as a painless lump. The treatment for an infected chalazion is the application of hot compresses (using a clean washcloth and hot tap water) and antibiotic drops or ointment. This will make the infection subside, but the cyst itself will go away only in rare cases.

Once the acute infection has cleared up, the chalazion will probably have to be removed surgically. This is a minor operation that can be done with local anesthetic in an eye doctor's office, and it leaves no noticeable scar. Although this is never an emergency operation and there is no danger that the cyst will become malignant, I usually recommend that chalazions be removed since they tend to become reinfected.

Recurrent chalazions can be caused by an undetected chronic eyelid infection or by muscle pressure on the eyelid glands. Chalazions are not contagious.

Subconjunctival Hemorrhage

A common and frightening-looking eye problem that is not at all serious is a burst blood vessel in the eye. It is called a *subconjunctival hemorrhage* and is really just like a bruise anywhere else on the body. One of the tiny blood vessels in the eye bursts and blood seeps between the conjunctiva and the sclera. Because the conjunctiva is clear, the blood looks bright red instead of black and blue as it does in a bruise. A subconjunctival hemorrhage may cause a large red area in the eye, but because it is completely painless, it is most commonly noticed by someone else.

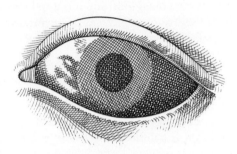

A subconjunctival hemorrhage.

The hemorrhage may be caused by trauma to the eye, but it usually happens completely spontaneously, and there is no treatment for it. The blood will be reabsorbed, gradually, in the course of ten to fourteen days. There is nothing effective you or your doctor can do to make it go away faster, but don't worry about it. This condition does not suggest an underlying illness of the eye or any sort of circulatory problem. You will probably want to see your eye doctor anyway, since it does look so scary, and he will probably assure you that it is just a burst blood vessel and nothing to worry about at all.

Eyelid Tics

Another problem that patients often complain to me about is a twitching of the upper or lower eyelid. This is not a medical eye problem, although it can be a nuisance because people feel self-conscious about the twitch and are certain others can notice it. Although it feels very obvious, the twitch is barely perceptible. It is caused by *fibrillation,* or quivering, of the muscles around the eye, invariably the result of tension or anxiety, and not any kind of eye disease or eye muscle problem. I find that telling my patients to try to ignore the twitch and explaining to them that tension is at the bottom of it often clears up the problem.

Although there are a number of other external eye problems, the ones I have described above are the only really common ones. Many cause discomfort to the eyes, and most of them should be treated medically, so you shouldn't feel encouraged to diagnose and treat them at home. As you can see, the symptoms are similar in many cases, and the treatments often require medication that only your doctor can prescribe. But understanding the nature, causes, and treatments of these eye problems should help alleviate any anxiety you may feel when you or a member of your family experiences external eye discomfort.

6

Childhood Eye Problems

MYTH: Rolling or crossing your eyes a lot is a bad idea because it can make you cross-eyed.

FACT: Although many of the causes of crossed eyes are not well understood, there is no question that you cannot cause it by rolling, crossing, or in any other way using your eyes or external eye muscles. These muscles are meant to be used, and there is really no way to misuse them. Rolling your eyes is not very attractive and it may make your eyes tired, but it is not at all harmful.

MYTH: Children should have their first eye examination when they enter first grade.

FACT: Every child should have his or her eyes examined by a medical eye doctor at age three. If there are no problems at that time, the next exam can take place at age six, but if the doctor spots a problem that might interfere with the child's learning to see, it can be corrected while the child is still young enough to develop normal vision.

MYTH: There is no need to worry about crossed eyes in young children because they will grow out of it.

FACT: Crossed eyes are normal during the first twelve to eighteen months of life, after which time a child's eye muscles and binocular abilities should be developed enough to permit parallel alignment of the eyes. A child beyond that age who is cross-eyed will not, in all likelihood, grow out of the tendency, and this is indeed something to worry about.

Young children are vulnerable to many of the same optical defects and eye disorders that adults are. These and other conditions that are exclusive to or that generally first show up in early childhood present a particular problem, since it is during that time that children learn to see.

The two main components of normal vision are 20/20 eyesight and the ability to use the two eyes together *(binocular vision)*. Nature has provided the first six years of life for learning these two skills; after six, or so, it is no longer possible for a child to pick them up. Any interference with this learning process because a one or both of a child's eyes are less than healthy and structurally normal can seriously and permanently impair the development of normal vision. Thus it is vital to correct eye problems in children as early as possible—the longer a child has to learn normal vision the better. Early examination of your child's eyes is, therefore, a major parental responsibility.

The two conditions I will discuss in greatest detail in this chapter are of such importance because, as common as they are, many parents do not understand their nature and consequences. I want to inform you of the facts about them so you will not be guided by old wives' tales and dubious practices when it comes to looking after your child's eye health. In most cases, early diagnosis and rather simple treatment can correct children's visual defects and set them on the road to learning normal sight. The keys, however, are early diagnosis and a willingness on the part of parents to be supportive and reassuring, rather than clinging to false hopes that their child will grow out of the problem without needing treatment.

Why Childhood Eye Problems Cause Visual Disabilities

Binocular vision, the ability to use both eyes together and deliver two images of the same object to the brain, is a characteristic of man and the higher mammals. Lower animals, such as birds, fish, and reptiles, have only monocular use and never see the same object with both eyes at the same time. Rather, each eye sees something different and delivers a different message to the brain, which then selects one message and suppresses the other. Binocular vision is what is responsible for depth perception. Lower animals and, for that matter, people who have only monocular use, are not totally without a sense of three dimensions, although their perception of it is less vivid and they must derive it from other, nonvisual clues.

In order to develop binocular use, we must be able to hold our eyes parallel. If the object we are looking at is straight ahead, both eyes must look straight ahead; if the object viewed is to the right, both eyes must move to the right and remain absolutely parallel. If they do not, the brain gets two different images. The brain of a child under six would respond to this by suppressing the image from the eye that is off course in order to avoid the extreme annoyance of double vision. This may sound like an advantage, but it is not. Suppression is learned just as binocular use is, but it is impossible to learn both. If the problem is not corrected in time, this child will never learn to use both eyes together, and in some cases will not be able to learn 20/20 vision either.

The second major component of normal vision, expressed as 20/20 visual acuity, depends first on an eye without optical defect, but the anatomical capacity is not all that is needed. Remember that the macula is the only part of the retina that can deliver a 20/20 message to the brain and that all other parts of the retina send images that are less clear. When both eyes are parallel to each other, the object viewed is focused on both maculas and not on a peripheral part of the retina in one eye. If a peripheral part of the retina of one eye is receiving an image of the object viewed, the macula of that eye is getting a clear image of something else. The brain of a child under six will suppress that second message, which prevents the transmission of a message from the macula to the brain, and this, in turn, prevents the development and learning of the connection between that eye and the brain. When both eyes

Clear central vision (20/20) is received by the macula. Less clear peripheral vision is received by the rest of the retina.

send a clear macular message of the same image, the brain has a chance to learn visual acuity in both eyes, but the lesson must be learned in the first six years of life.

Children as Eye Patients

It is a particularly unfortunate fact that eye problems in children are frequently not corrected, either because parents believe their child will simply grow out of them or because they doubt their young child can cooperate adequately with a complete eye examination. But children will not usually grow out of these problems, and children as young as three years old can certainly undergo a routine eye exam.

That examination, however, must be done by an eye doctor, not the child's pediatrician. Many parents assume that their child's doctor, in performing a few simple eye tests, has taken adequate care of the matter. Although the most obvious eye problems can be picked up when a pediatrician looks into the child's eyes and has the child read a simplified vision chart, there is a broad range of problems that cannot be spotted in this way. For one thing, examination of an undilated eye is limited. For another, the need for glasses, particularly if it is in one eye only, cannot be determined by a mere reading of an eye chart. And, perhaps most important, a pediatrician does not perform any tests that would reveal a tendency toward crossed eyes if it is not clearly obvious. In this particular case —one of the most significant in your child's visual development—a pediatrician is little more able than you are to spot an important eye problem.

It has been the common practice to schedule a child's first visit to the eye doctor at about age six. This potentially dangerous practice has permitted unnecessary but serious and irreparable visual handicaps to develop. An eye problem that begins in these early years is usually easy to diagnose and correct if an eye doctor is consulted. The ideal age for the first routine visit is at three years. Then any necessary correction can be made while there is still ample time for the child to learn how to see properly. By age six it is usually too late.

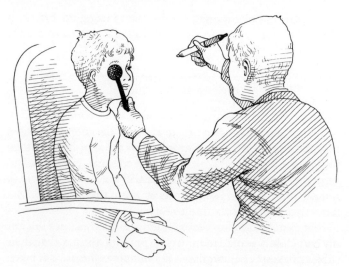

A three-year-old child can undergo a complete eye examination.

Let me tell you about the procedure I use for examining the eyes of a young child. In general, I perform the same examination on children as I do on adults, except that I do not test for glaucoma and other "middle-age" eye problems. The main hurdle I seek to overcome is children's fear of a stranger who seems to command a host of unfamiliar and frightening instruments. I have found that a friendly approach, coupled with a few games to put the child at ease, will most often solve this problem. I also ask that one or both parents be present if at all possible. Various instruments can be used as toys—I let children play with the flashlight I will later shine in their eyes, and I use pictures and vision charts to tell stories. In my office the examination chair can be raised and lowered, and I often use it to give my young patients rides.

The basic preventive eye examination is a relatively short procedure—fifteen to twenty minutes usually suffice—and none of the tests causes any pain. Dilating drops do cause a slight degree of discomfort, but the effect is temporary. I try to alleviate children's anxieties about the drops by assuring them that the stinging sensation lasts only a few seconds. I have found that the temporary blur they cause does not usually bother children.

In the case of a very uncooperative or frightened child, I use two visits for the examination. During the first, games and a generally relaxed atmosphere put the child at ease, and a few simple tests can be tried. I then give the parent drops and instructions in their use so they can be put in at home just before the second visit, the fearful anticipation of which will be less since the first visit was a pleasant experience.

I need only a small degree of cooperation and response from my patients to do an accurate and complete eye examination, and it is possible to conduct a very thorough one with no response whatever. Very young children, senile adults, or retarded or brain-damaged patients, all of whom may be unable to participate actively in certain tests, can be examined quite successfully. The major part of the eye exam is based on my own observations, and since information from my patients' subjective responses overlaps the objective data in most areas, it serves largely as confirmation.

In the past it was considered difficult to test the visual acuity of preschool children, since reading the letters on the eye chart was often beyond them. Various shapes, drawings of familiar objects, or a large letter E turned in various directions were often substituted for the regulation eye chart letters. Nowadays the widespread influence of *Sesame Street* has resulted in scores of three-year-olds who can cheerfully rattle off the alphabet and will enthusiastically read the random letters on the eye chart.

If all is well, the next exam need not take place until age six, but if any problems are identified, they should be corrected while it is still possible for normal vision to be learned. There is, therefore, no reason that a three-year-old cannot have a complete eye exam, especially since this exam is the only way to ensure that a child has two normal and healthy eyes with which to learn the visual skills acquirable only during early childhood.

Let us now consider the two major groups of eye problems that develop in early childhood and that can, and must, be diagnosed early if normal visual development is to take place.

Amblyopia and strabismus are the two main developmental conditions we look for in young children. Because they most commonly make their appearance during the time when a child is learning to see, their main effect is to interfere with visual development. What makes amblyopia and strabismus so serious is that, if they are not corrected in early childhood, they can cause lifelong visual handicaps.

Amblyopia

Amblyopia is the medical term for an anatomically healthy eye that has not learned how to see properly because of some uncorrected defect. A disease within the eye can also prevent normal visual development, but this is not called amblyopia. The common name for amblyopia is *lazy eye,* a misnomer because it suggests a weakness of the eye muscles or an eye that does not have the strength to focus on objects straight ahead. In fact, an amblyopic eye has all the anatomical potential to see normally, but it does not do so simply because the brain will not learn to interpret an inadequate visual message.

Many things can cause amblyopia, among them such optical defects as moderately high astigmatism in one eye. For example, a child's right eye is optically normal and transmits a sharp visual image to the brain, which responds by developing normal visual acuity in that eye. The astigmatic left eye receives a blurred image on the retina and cannot send the brain a better image than it receives, so the brain has no chance to interpret 20/20 vision from that eye. If the child passes age six before the astigmatism is corrected, the brain will never have had the chance to learn normal vision through that eye and it will not be able to pick up on its lessons at a later date. The right eye will have learned how to see 20/20, but the left eye will not have had a chance to learn normally. Depending on the degree of astigmatism, the left eye will have learned to see only as acutely as the uncorrected astigmatism allows. Other optical defects—farsightedness or high nearsightedness —can have the same effect.

If the optical defect is extreme enough, in addition to being unable to develop visual acuity in the amblyopic eye, the child may be unable to learn to use both eyes together normally. Some amblyopic individuals have reduced depth perception, but they can compensate for this by using such nonvisual clues as touch and sound, as well as the learned knowledge and memory of where objects should be in relation to each other. Their world will not be flat, but it will not be as vividly three-dimensional as the world of people who have developed binocular vision in two 20/20 eyes.

Another possible cause of amblyopia is uncorrected *strabismus,* or crossed eyes. There is a cause-and-effect relationship between the two conditions, but it is not exclusive. Crossed eyes can cause amblyopia but does not always; similarly, strabismus can result from amblyopia as well as from other conditions. For example, in one type of strabismus, one eye looks straight ahead along the visual axis, but the other eye is always turned inward. The straight, or *fixing eye,* receives and sends a clear visual message of what it sees to the brain. It will develop normal visual acuity and be able to see 20/20 by age six. The turned in, or *deviant eye,* will receive a clear image of what it is looking at, but because that is something different from what the fixing eye sees, the visual message it sends to the brain would result in double vision. The brain of a young child deals with this extreme annoyance by learning to suppress the image seen by the deviant eye. If the images continue to be different through age six, the brain will continue to suppress, and the deviant eye, unable to learn because suppression has deprived it of all chance to learn to see, will never be able to interpret sharp central images. The result is permanent amblyopia in that eye.

Although the strabismus that causes this might be obvious to a parent, it could just as likely be too slight to be noticed, but great enough to cause suppression and amblyopia. An eye doctor can readily spot the problem with a few simple tests and solve it by correcting the inward tendency of the deviant eye. If corrected early, that eye will have time to learn to see properly, but if left uncorrected until after age six, any correction would be for appearance only since the deviant eye would be permanently amblyopic and no longer able to learn seeing.

Does that mean that a normal-seeing adult who for some reason develops an inward turning eye will become amblyopic? No, it does not. If the two eyes of an adult are forced to look at different things—if, for example, an eye injury has damaged the nerve or muscle connections of one eye—the adult *will* see double because the brain will not be able to suppress the deviant image. Suppression also must be learned before age six, so it is too late for an adult to avoid double vision in this way. This condition is usually correctable by special glasses or surgery, or if that does not do the trick, a patch on the deviant eye will prevent double vision.

Regardless of cause, constant suppression is all that is required to make an eye amblyopic. The eye can be totally without optical defect, but the failure of the brain and eye to establish communications will thwart the development of normal vision in that eye. For example, if a child with perfectly normal and healthy eyes were to have one eye patched at birth and the patch left in place until age six, that eye would be amblyopic when the patch was removed and would remain so throughout life. No matter how many opportunities the eye had to see after the patch was removed, the time for learning that skill would have passed.

How is amblyopia diagnosed? Children will rarely complain of blurred vision or less than vivid three-dimensional perception since they know of no other way of seeing. And if strabismus is the cause, it may not be clearly visible. How, then, can I tell that vision is not developing normally? In the course of the routine eye examination, I perform a series of short tests that show if both eyes are being used and if each eye sees what it should. I can also identify what has caused the amblyopia—information I need to know so I can devise treatment for it.

In all cases, the treatment for amblyopia is correction of the factor that causes it. Children under six can pick up on the learning process once their learning tools are in proper working order. This often means glasses will be prescribed, and this is something many parents find it difficult to deal with. No one likes to think that his or her child has any sort of physical problem, and the idea of a young child with glasses seems to many to be an advertisement of this fact. I would not pretend that this is not a justifiable concern, but I would like to point out that parents must make a choice, on their child's behalf, that can affect the entire course of their child's life.

If your eye doctor tells you your child needs glasses, think before you gasp: would you prefer glasses now and the normal visual development they will allow, or would you rather that your child went without them in early childhood and perhaps was unable to be an athlete, an airplane pilot, or whatever he or she wanted to be, because of a permanent visual handicap? Vanity about glasses is rarely experienced as intensely in children as it is in adults. Children, particularly of the age we are talking about, are eager to see well and take their place with their playmates in the various activities that require good and normal vision. Although pre-teens and teens might feel self-conscious about glasses, I have rarely encountered an under-six child who was reluctant to wear them, unless he or she had been influenced by a parent into believing that there was something to be ashamed of.

Strabismus

Strabismus, or crossed eyes, is the second of the common childhood onset eye problems. Like amblyopia, it can seriously impair the visual learning process if uncorrected before age six. Strabismus describes two eyes that are not perfectly parallel when viewing an object. This does not mean that the eyes have to be straight ahead; they simply must be parallel to each other whichever way they are turned. But like lazy eye, "crossed eyes" is a misnomer. Although it is possible in one type of strabismus for the lines of sight (visual axes) to cross, they are not always crossed and certainly at no time do the eyes themselves cross. Some other common names for strabismus are "a cast to the eye" and "wall eyes," but these terms are even less correct than "crossed eyes."

Although strabismus is often very obvious, it is frequently impossible to spot with the naked eye. All the same, it is no more possible to be a little bit cross-eyed than it is to be a little bit pregnant. Any degree of strabismus will have the same visual effect; whether it is a slight or major deviation, the damage done to vision is the same. A particularly unfortunate bit of misinformation that contributes to the number of children who are seriously and permanently handicapped by strabismus is the notion that they will grow out of a tendency to cross their eyes. Although it is true that a certain amount of random divergence or convergence is common in infants, children past the age of one or one and a half should be able to hold both eyes in alignment. Crossed eyes after that age is not normal and cannot be left to improve on their own.

By the same token, you cannot give yourself strabismus. The often-heard warning that rolling your eyes or crossing them in play might make you permanently cross-eyed is completely fanciful. Your external eye muscles are meant to be used, and they are designed to move your eyes in all directions, as well as to hold them parallel to one another. You cannot misuse or overuse these muscles.

Like amblyopia, strabismus is damaging because the brain is constantly given an unacceptable visual message and that interferes with the development of visual skills. Use of the two eyes together is impossible since they are viewing different things. Never having had the chance to receive two similar messages, the brain is unable to learn to assemble a three-dimensional image. Without this learned skill, the individual will never have normal depth perception. And, of course, the deviant eye can become amblyopic.

There are several possible causes of strabismus, some better understood than others. And in some instances we cannot identify the cause at all. The most obvious one—that the eye muscles themselves are too weak to hold the eye in alignment —happens to be relatively uncommon. There is no question that there is a hereditary influence; children whose families have a history of strabismus will have a greater tendency to develop it. Another possible cause is a malfunction of the nerve connection to the external eye muscles. A common, more readily explainable cause is an uncorrected high degree of farsightedness. Remember that young people can correct farsightedness by using their near-focusing ability. This involuntary action plays a part in strabismus because when near-focusing muscles are used the eyes automatically converge to take in the near object. Notice that when you shift from looking at a far object to looking at a near one your eyes turn in a bit. This is a natural and normal reflex, but when a farsighted child uses the near-focusing muscles to view far objects clearly, his or her eyes may converge. The reflex is stronger in some than in others, so it does not mean that all children with uncorrected farsightedness will develop strabismus, but it is a possibility.

A disease that causes poor vision in one eye is another possible cause. If one eye sees quite badly, there is not much visual benefit to be gained from using it. The brain will not tell the nerves to tell the external muscles to hold the eyes parallel, and the defective eye may simply turn in or out because there is little reason for it to hold itself parallel to the other eye.

What difference does all this make? Is strabismus more than

just a cosmetic problem, a matter of looking a bit odd because the eyes are crossed? Indeed it is. Binocular use is not a skill mechanically achieved; it must be learned in that ever-important period before age six. If strabismus is uncorrected during that time, the child will never be able to learn to use both eyes together. Correction after age six will improve appearance, but it cannot provide a second chance to learn binocularity. Likewise, if the strabismus has caused the deviant eye to become amblyopic, correction of the deviance after age six will not cure the amblyopia.

Extremely misaligned eyes can be spotted by a parent or anyone looking at the child, but strabismus is often not that pronounced. An eye doctor, however, can readily discover strabismus during the routine eye examination and can determine how it should be treated.

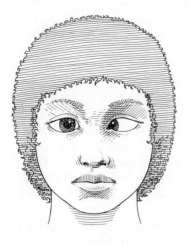

Strabismus.

Epicanthus.

A condition called *epicanthus* often causes parents of young children to suspect strabismus. At birth, a wide nose bridge normal to all babies is combined with an unusual eyelid fold that often makes it seem that one eye is turned in too far when the baby looks slightly to one side. In fact, a large portion of the sclera is hidden by the *epicanthal fold,* but the eye is not turned in more than normal. This is an anatomical feature that is not at all related to strabismus, and it does not interfere with learning to see. The child may look abnormal, but he or she does not see abnormally. And in most cases, the epicanthus recedes as the child's nose narrows.

I have found in my practice that the thing that frightens parents most and makes them want to avoid finding out for sure if their child is cross-eyed is the thought of surgery to treat the condition. But is surgery always necessary? No, not at all. We always try the simplest approach that will do the job, and there are a number of treatments that can be tried before we resort to surgery.

Glasses are the simplest treatment for strabismus. If the cause is high farsightedness, glasses can be prescribed to correct it. Once a child no longer uses the near-focusing muscles to correct the farsightedness, the convergence reflex is not triggered and that alone may make the strabismus disappear. Sometimes special bifocal glasses, which would otherwise never be used for young people, are worn to minimize the normal convergence when near objects are viewed. In cases where a child cannot be made to wear glasses, a certain kind of eye drop can be tried to inhibit the reflex. These drops do not, in any circumstances, correct farsightedness, and they are not so safe or effective as corrective lenses, but they can be used to benefit those rare cases of children who absolutely refuse to wear their glasses. Of course, as the child grows older and becomes more amenable to glasses, he or she will have to wear them to correct the farsightedness. I am personally reluctant to use these drops because I believe glasses are safer than is constant administration of medication.

But what if the strabismus is not caused by a correctable optical defect? Can eye exercises be used to force the eyes on a parallel course? Although one type of eye exercise can be useful when there is a convergence weakness when looking at near objects, the kind of exercises most commonly used for strabismus have little practical application. These so-called orthoptic exercises aim to stimulate parallelism and binocular use by asking a child to bring together two disparate images, usually combining two pictures, such as a bird and a birdcage so the bird appears to be in the cage. The trouble with this is that a child old enough to understand the exercises and repeat them as many times and for as long as would be required to "learn" the lesson is usually too old to get any benefit from them. It is the rare under-six child who could perform the exercises as prescribed, and the majority of children under six who are involved in these courses are being made to use up precious time in their key learning years. Parents who are reluctant to accept their child's need for glasses are particularly attracted to the idea of orthoptic exercises, but the trade-off is expensive, time-consuming, and usually unproductive.

There are isolated cases in which specific exercises might be helpful, but only your eye doctor can tell you if your child could benefit from them, and if so, when and where your child should go for treatment. Do not involve your child in a program of eye exercises without consulting your eye doctor.

Eye patching is another commonly seen approach to treating strabismus, but many people do not realize that patching is always used in combination with glasses or surgery, never as a treatment in itself. When the deviant eye is also amblyopic, a patch on the good eye can stimulate the amblyopic eye to learn visual acuity so that when corrected by glasses or surgery that eye will have the incentive of visual reward to hold itself parallel to the other eye. If a normal eye and a poorly seeing eye are straightened without patching beforehand, it is likely that the poorly seeing eye will go off course again because it still does not see as well as the other eye and, therefore, has little motivation to stay parallel. If the good eye has been patched, however, and the poorer eye given a chance to develop good vision, once it is straightened it will immediately have sufficient visual reward to keep itself on course with the other eye. A patch alone will not cure strabismus, but it is an effective tool when used in conjunction with glasses or surgery.

Strabismus surgery is the last resort if optical correction of the causative factor does not work or if the cause is not an optical defect or is not known. Surgery is generally quite successful, but more than one operation is sometimes necessary. The surgery can accomplish two possible ends. The first is to permit the eyes to develop normally once the errant eye is on a parallel course with the other eye. The second aim is cosmetic—realigning the eyes will improve the appearance. If surgery is done after age six, the cosmetic improvement is the only one possible. The person will look normal, but will not be able to get any visual benefit from the surgery. The previously deviant eye will either have no developed vision and be amblyopic, or both eyes will be able to deliver a visual message to the brain but the brain will not be able to assemble a three-dimensional image, since by then it is too late to learn binocularity.

Strabismus surgery is safe, requires a very short hospital stay, and poses no threat to vision, since it is performed on the external eye muscles and the eyeball itself is never entered. The psychological trauma of surgery, a significant concern as the patients are generally young children, is quite minimal. Patching after the operation is usually just for one day, during which time the child is too groggy from the anesthetic to be bothered very much by the patch. Some doctors do not patch after the operation at all. It is often possible for a parent to share a room with the child and be available to ease the child's anxieties.

I strongly suggest that the surgical procedure be explained to children in as much detail as they are able to understand, since the fantasy of the unknown is always more frightening than a basic understanding of an unpleasant experience to come. No child should go into the hospital fearing that his or her eye will be removed from its socket or that the eyeball itself will be cut into, though I have found that parents are worried about this more often than are children.

The operation itself involves altering the strength of the external eye muscles in such a way that the eyes will be parallel. There are a number of different ways to do this, so your doctor will determine, on the basis of careful pre-operative measurements, the approach he feels will achieve the most predictable result. But you should realize that surgery is performed on people, not machines, and it is often difficult to estimate exactly how much correction will be gained through the chosen procedure. It is, therefore, sometimes necessary to do a second operation after the results of the first have become clear. This does not mean that the operation did not work or that the doctor made a mistake. It simply means that the individual strength of any child's eye muscles is subject to many variables that are difficult to determine precisely in advance. It is a good idea to keep this in mind if your child is scheduled for strabismus surgery so you will not be upset or discouraged if your doctor says a second operation is needed.

Visual Training for Learning Problems of School Children

I would like to take a moment here to discuss a recent and unfortunate trend that has become something of a national mania: the referral of school children to "visual training centers" by teachers and other school personnel who, in well-meaning ignorance, attribute early reading and writing difficulties to eye problems that could be solved by "visual training." The supposed difficulty usually focused on is the very common tendency of children learning to read and write to place letters and numbers backwards, upside down, or in reverse order. This does not represent a visual problem. For one thing, eight- and nine-year-olds (the age most frequently in question) have already passed the time when they learn to see, and for another, it is a simple fact that all children tend to write letters and numbers in various incorrect ways during the time when they are first becoming familiar with these new processes. This is perfectly normal, and in most cases the child will eventually correct himself. Learning to read and write is like learning to walk—a toddler will trip over his or her feet twenty times and on the twenty-first try will learn to put the left foot forward and follow it with the right. If you would not send your child to walking school because the first attempts at walking met with a few stumbles, why send a beginning reader for "visual training"? It is important to realize that learning to read and write involves a barrage of unfamiliar information, symbols quite different from those previously used by children, and that it is the strangeness, rather than any sort of eye abnormality, that is causing confusion.

The so-called visual training centers, which are staffed exclusively by optometrists and nonmedical technicians, can do absolutely nothing to correct the so-called problem or hasten learning to read and write. You may have heard a glowing report or two from parents whose children have had visual training, but this is invariably because the training coincided with the period when the child would, as a matter of course, have begun to write properly. In this way, this expensive and time-consuming process feeds on itself, garnering credit that properly belongs to the passage of time rather than to any virtue of the program.

This is not to say that there is no such thing as a learning disability, though I sometimes wonder if dyslexia and the other fancy names for school problems that are bandied about are much more than an extension of Newspeak. Optical defects or neurological or psychological problems can cause a child to have a difficult time in school, but these are hardly common, and they are matters for diagnosis and treatment by eye doctors, neurologists, or psychiatrists. If your child's school has indicated a suspected "learning problem" and recommends visual training, walk quickly in the opposite direction—to your eye doctor. He can examine your child's eyes and tell you if the problem is a real one, if your child is simply going through the normal period of visual orientation, or if, in the absence of any eye problem, consultation with a neurologist or psychiatrist is recommended. Do not, in any circumstances, accept the diagnosis of a person who is not trained in eye medicine, or follow such a person's prescription for treatment by a second nonmedical practitioner who is unable to correct the supposed problem in any way.

Amblyopia and strabismus, which originate in the period before age six, are really the only frequently seen eye problems that are exclusive to young children. The treatment of these and of other eye problems that children have in common with adults is in the domain of your eye doctor. But whether corrected by glasses, surgery, or a combination of the two, strabismus and amblyopia are relatively easy to treat. If correction takes place before age six, a child can look forward to a life of normal vision, since neither strabismus nor amblyopia make a person more likely to suffer other eye problems. But if correction is not undertaken at this early age—whether because of ignorance of the condition, fear of surgery, useless attempts at correction with eye exercises, or the mistaken belief that the child will grow out of the problem—the strabismic or amblyopic child will grow up to be a strabismic or amblyopic adult who can never have binocular vision or fully developed visual acuity in both eyes. This is a tragic and unnecessary deprivation, since early diagnosis would have opened the way for correction and unimpaired learning of visual skills. That is why it is absolutely essential that all children have their eyes checked before they outgrow their ability to learn normal vision. The examination at age three is, therefore, perhaps the most important visit to an eye doctor anyone can make. No child should be prevented from having it; the consequences could be with him or her throughout life.

7

Retinal Disorders

MYTH: It is a good idea to eat carrots and other sources of vitamin A because they improve your ability to see in the dark.
FACT: Although it is true that photochemical changes take place in the retina to permit vision in the dark and that the retina uses vitamin A for this purpose, the amount of vitamin A you get in a normal, balanced diet is more than sufficient for this and all other body needs. Supplementing this with large doses of the vitamin will not improve your vision, in the dark or elsewhere. Your body simply cannot use the excess, which in large quantity can even be harmful.

MYTH: If you cannot see well in the dark, you have night blindness, which is a common problem.
FACT: Night blindness is not at all common; in fact true night blindness is a symptom of a rare condition called *retinitis pigmentosa.* Most people have more difficulty seeing at night simply because it is harder to see well when there is less light. This is not night blindness.

MYTH: People who are color blind see only in black and white.
FACT: Although dogs and cats and many other animals see in black and white, not even color-blind people see no colors at all. Color-blind people perceive colors less vividly than nor-

mal-seeing people, but their world is never completely mono-chromatic.

The retina is the delicate, thin structure that I have compared to the film of a camera. It is responsible for receiving visual messages—what we see—and sending them to the brain by way of the optic nerve. Your ability to see colors is part of the function of your retina, as is your ability to see well in different levels of light. In order to do its work, the retina should not be scarred or torn, and nothing should interfere with the work of its tiny blood vessels. No blood should be present anywhere in the retina but in those blood vessels. Furthermore, the adhesion of the retina to the two structures that lie on either side of it—the choroid and the vitreous fluid—must not be interrupted.

When any one of these conditions is not fulfilled, vision can be affected. The seriousness of the effect depends on where on

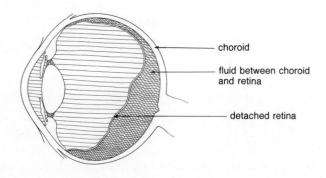

Retinal detachment.

the retina the problem exists. If it is out on the far edges—that part of the retina that receives peripheral visual information —it is unlikely that there will be any noticeable loss of vision. The closer to the center the problem gets, the more noticeable it will be. And any damage to the macular area—the specialized area of the retina that provides sharp, central vision—will be very noticeable indeed.

Retinal Detachment

One of the most serious retinal problems is referred to as *detachment*. This means that something has interfered with the adherence of the retina to the choroid and the retina has pulled away. The possible causes for this are numerous, some known and others unknown. Torn retinal tissue can cause detachment, or detachment itself can cause the retina to tear. Other causes can be a blow to the eye, a cyst or tumor, scar tissue, a hemorrhage, infection or some other disease of the eye, or simply a general tendency toward poor adhesion. In very rare cases severe nearsightedness can predispose to detachment.

A detached retina does not hurt. Symptoms include visual loss (especially in one area of the visual field), spots before the eyes, and light flashes. But this does not mean that anyone who experiences these symptoms has a detached retina. If you notice such problems with your eyes, see your eye doctor as soon as possible, so that he can identify the cause and begin appropriate treatment, if necessary.

Retinal detachment is serious. For one thing, it will nearly always get worse if it is not treated quickly. The detached area tends to get larger, or the retina may tear away in several places. Wherever the detachment occurs, the retinal tissue no longer functions as it should and some sight is lost. The treatment of retinal detachment is always surgical. The aim is to reattach the retina and do everything possible to make sure it does not detach again. When the underlying cause is known, it, too, can be treated. But often the cause is not known or, as in severe nearsightedness, is incurable.

There are a number of ways to repair a retinal detachment, none of which leaves a visible scar or requires cutting into the eye itself. Retinal repair can be done by burning around (or cauterizing) the tear to seal it and reattach the retina or by using an ice probe (freezing) to accomplish the same end, but these procedures are done on the white of the eye. Sometimes lasers are used to cauterize the tear, and that is done by aiming the laser beam through the pupil. Another procedure shortens the eyeball slightly by taking a tuck in the back portion of the eyeball and securing it with silicone. Some of the external eye muscles may be cut to give access to the part of the eye being repaired, but they are reattached after the repair has been done.

Unfortunately, the success rate of retinal surgery is not so high as it is with other eye procedures. When it is a success, the retina stays reattached and vision will be unimpaired. Sometimes the retina stays reattached but damage has already been done to the retinal tissue and some vision is lost in the damaged area. The degree of the impairment of vision depends upon where the detachment took place. If it was on the periphery, little if any visual handicap will result; if it was closer to the center, a greater loss of vision can be expected. In some cases the retina may become detached again. What's frustrating about this is that, when we can't identify or treat the underlying cause, the possibility of a recurrence is increased.

But before the picture gets too gloomy, let me say that I have mentioned retinal detachment because it is serious and dramatic, not because it is common. We occasionally hear about athletes who suffer retinal detachments because of some sports-related injury, but by no means should you be afraid that you or a member of your family will suffer it in the normal course of events. It simply is not a common occurrence.

The vast majority of people who experience symptoms of spots before their eyes are suffering not from retinal detachment at all but from what we call *vitreous floaters.* These small conglomerations of cells have been released from the lining of the inside of the eyeball and literally float in the vitreous. They are usually apparent when one looks at a white background in certain light levels, and they appear as tiny spots, specks, or doughnut-shaped forms. Vitreous floaters are not a disease, though they can be a nuisance. They tend to occur more in nearsighted people and, although they can be recurrent, they most often are reabsorbed and disappear on their own. There is no effective treatment for vitreous floaters, and the best thing to do is to try to ignore the minor annoyance they can cause.

There are a number of other conditions that affect the retina. Some can cause detachment or other, less serious problems. Some can be treated, some go away by themselves, and some can be left alone.

Retinal Hemorrhage

A hemorrhage is simply the term for a break in a blood vessel. This can occur in various parts of the body and can be caused by a number of different things. The problem when it happens in the retina is that the blood interferes with the function of the retinal cells that are receiving visual stimuli. In most cases a hemorrhage will take care of itself; the break in the blood vessel will seal and the blood will be reabsorbed and carried away as waste. In that case, partial loss of vision, if indeed there is any, is usually temporary. In other cases, the hemorrhage may turn to fibrous scar tissue. Then the work of the retinal tissue will continue to be interfered with and that part of the retina will not be able to receive or send visual information. Depending on where the scar is, the loss of vision may be serious, minor, or unnoticeable in any practical way. The most serious danger of a retinal hemorrhage is if the scar tissue pulls at the retina and causes it to detach. Although they can have serious consequences, retinal hemorrhages are often quite minor. I have examined patients who show scars from past hemorrhages but who never knew they had them because they have never experienced interference with their vision.

Retinal Infections

Like all other tissues in the body, the retina is susceptible to infection. The source of infection is usually somewhere else in the body, and the blood carries the infectious agents to the retina. Such varied conditions as tuberculosis, parasites, syphilis, and viruses can infect the retina. The best treatment is to cure the underlying infection. If, while examining your eyes, your eye doctor notices an infection you may not be aware of, he may refer you to your family doctor. Your eye doctor and family doctor can then work together to try to discover the underlying infection. In the best circumstances, the infection can be cleared up and you will experience no loss of vision. In some cases, retinal tissue in certain areas will be damaged and no longer function. And, in very serious cases, the scarring caused by the infection could bring on a retinal detachment.

Other Retinal Disorders

A blow to the eye can cause injury to the retina, provoking a hemorrhage, swelling, or even detachment. In most cases, the injury is minor and temporary, but it can be serious. If you are struck in the eye and notice blurring or spots before your eyes or partial loss of vision, you should see your doctor right away.

Much rarer retinal problems are tumors and cysts, but these are usually benign. Depending on their size and location on the retina, growths can cause partial loss of vision and detachment. Treatment will be determined by the cause, location, and character of the tumor or cyst.

Color Blindness

Color blindness is an untreatable condition of the retina that is not very well understood. There is something wrong with the specialized cells of the retina that are responsible for the perception of color, but the exact problem is unknown. We do know that color blindness is hereditary and occurs almost exclusively in men, although it can be genetically transmitted through women. Color blindness occurs at birth and never develops later in life, nor can it get better or worse as one grows older.

Color blindness can be slight or it can be severe. In general, though, the color-blind person can adjust to the problem and "perceive" colors from other visual clues. In no case does a color-blind person see in black and white. Usually, he will simply have a less vivid perception of a certain color or colors and be able to perceive a narrower range of shadings. Obviously, color blindness is a serious handicap for someone who works in the visual arts or other fields that require exact and specific perception of colors, but for most color-blind people, it is a minor handicap easily lived with.

The Retina and Circulatory Diseases

One of the nicest things about the retina is that it is very easy for your eye doctor to examine it. In the course of the routine eye examination, he will dilate your pupil and look inside your eye, directly at your retina. And because the retina is the only place in the whole body where living blood vessels are visible for direct examination, when your eye doctor looks at your retina he can see how well your circulatory system is working. This is something that cannot be done with blood vessels elsewhere in the body, and it is important in diagnosing a particular group of circulatory ailments that also affects the retinal blood circulation. Although in all cases the treatment is aimed toward the condition as a whole rather than the retina alone, it is useful for us to know what these disorders are, how your eye doctor can diagnose them, and what effect they have on the retina and vision in general.

One of these conditions is *arteriosclerosis,* often referred to as "hardening of the arteries." This term is misleading since the arteries do not become hard but merely clogged with fatty deposits and are less elastic, providing narrower passageways for the blood. The overall effect of arteriosclerosis is less efficient circulation of the blood.

Under normal conditions, arteriosclerosis is a result of advancing age. As we grow older, we will all develop "hardening of the arteries" to some degree. But arteriosclerosis can also occur prematurely. There are various causes for this, among them diabetes and, many people believe, high-fat diets or at least an inability to metabolize fats properly.

What does all this have to do with the eye and the retina in particular? If the arteries throughout the body are becoming clogged, the arteries of the retina are no exception. Retinal circulation slows down, the delicate cells of the retina are not so effectively fed and washed by the blood, and the cells begin to die. When a retinal cell dies, one less cell is available to receive and transmit visual information. How much this affects vision depends again on the location of the dead cells. If they are on the far periphery of the retina, they correspond to peripheral vision, and any loss of sight will probably not be noticed. If, however, the cells are near the macula, sharp central vision can be decreased, and that will be very noticeable indeed. Arteriosclerosis is a progressive disease, causing increasing clogging and greater cell death. In severe cases visual loss may result.

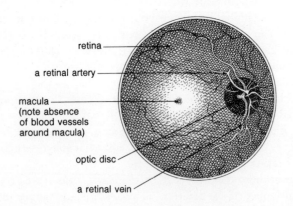

retina

a retinal artery

macula
(note absence
of blood vessels
around macula)

optic disc

a retinal vein

Retinal circulation.

The management of arteriosclerosis should be undertaken by a family doctor. There is no effective way for an eye doctor to treat the circulatory problem as it relates to the retina exclusively, but if one is for some reason unaware that he has arteriosclerosis, an eye doctor will be able to diagnose it during a regular checkup by looking at the circulation of the retina. He will then refer the patient to his family doctor.

Hypertension, or high blood pressure, is another condition that affects the entire body and can show up in the retinal circulation. The effect on the retina occurs only if the hypertension is severe and has not been treated until a later stage. When that has happened, hemorrhages may occur and can interfere with vision. Fibrous scar tissue, left by a hemorrhage, can pull at the retina and cause more hemorrhages, which can lead to additional scarring and eventual detachment.

A third serious and relatively common condition is *diabetes.* Diabetes is a metabolic disorder in which the body does not produce enough insulin to process the sugar it needs. Although it is incurable, diabetes can be controlled quite successfully with various drugs or insulin injections, and diet. Very serious, uncontrolled, or undetected cases of diabetes, however, do damage to the blood vessels of the entire body. One of the things that happens is that premature arteriosclerosis can result. A more serious effect is a condition called *diabetic retinopathy,* a cycle of retinal hemorrhages, followed by scarring, then abnormal proliferation of tiny blood vessels in the scar tissue, and more hemorrhages. Diabetic retinopathy is one of the leading causes of blindness today. Although an eye doctor will be able to say that the eye problem is the result of advanced or badly controlled diabetes, treatment must be aimed at the whole condition, not just at the retina. This is just another reason for careful control of the diabetes.

Most of us will never experience any eye problems related to the retina. But all of us should bear in mind that these problems, when they do arise, can be serious. It is fortunate that the retina lends itself so well to examination. All that remains is for you to make your retinas regularly available for that examination.

8

Cataracts

MYTH: Cataracts can grow back after surgery.
FACT: Cataracts are a clouding of the lens of the eye, and since the lens is removed in all types of cataract surgery, neither the cataract nor the lens can grow back.

MYTH: Cataracts can be surgically removed only when they are ripe.
FACT: A cataract is not a tomato and does not ripen like one. A cataract can get worse, but the time for surgery, if it is needed at all, is when the patient's vision is sufficiently impaired to interfere with his or her daily life.

MYTH: Older people frequently develop second sight.
FACT: This so-called second sight is usually a developing cataract, which produces a tendency toward nearsightedness. The increasing nearsightedness permits artificially good vision of close objects and frequently makes it possible to read without glasses even though they were previously needed. This is *not* a good thing, since it indicates a developing cataract.

MYTH: When you have cataract surgery, your eyeball is removed from the socket and then replaced after the operation is over.
FACT: No eye surgery involves removal of the eyeball from its socket. When cataract surgery is performed, an incision is

made in the front of the eyeball to give access to the lens of the eye. This is easily done without dislocating the eyeball.

MYTH: There is a new method of cataract surgery that uses laser beams.
FACT: Laser beams are never used in cataract surgery. People who believe they are have probably confused this with phako-emulsification, a cataract surgical technique that uses a needle-like instrument to suction out the lens of the eye.

The words *cataract* and *cataract surgery* inspire anxiety and wishful denial in most people, primarily because they are igno-rant of the facts about cataracts and eye surgery in general. It is an unavoidable though hardly pleasant fact that the majority of us, if we are fortunate enough to live to a ripe old age, will develop cataracts, for indeed the most common type of cata-ract is part of the natural process of aging. But it is also true that having a cataract does not automatically mean you need to undergo surgery. If surgery is needed, the cataract opera-tion is extremely safe and remarkably effective, and sight after surgery can be corrected to provide a far better picture of the world than was possible before the cataract was removed. Cataracts are a common concern, particularly to older people and their families, so let me try to shed some comforting light on the subject by explaining to you the nature, causes, and treatment of cataracts.

What Is a Cataract?

A cataract is any imperfection in the clarity of the lens of the eye for whatever reason. Because the lens must be perfectly clear to allow light to pass through it on the way to the retina, an imperfection of this sort can affect sight. How much effect it has depends on the size and density of the imperfection, and that can vary greatly. If you think of the lens as a window, a cataract can be like a speck of dust that is easy to look past and ignore, or it can be like a steamed-up window, which allows you to see only vague shapes and gradations in light and dark, or it can be like a window that has been painted black, which for all practical purposes makes it impossible to see anything.

Cataracts can affect vision in various ways: *(left)* vision with an early or minimal cataract can be close to normal; *(center)* vision as a cataract becomes more advanced can be fairly blurred; *(right)* vision with a severe cataract can be extremely blurred.

The most common cause of cataracts is simply increasing age; this is the so-called *senile cataract.* Because your lens grows throughout life, adding new cell layers to the outside periphery in a way similar to the growth rings of a tree, its size, resiliency, and clarity undergo change. The lens gets somewhat larger, but the major changes are in resiliency and clarity, since the layers are compacted and the lens tends to get more rigid and less transparent later in life. This happens to all of us, although the rate of change obviously varies among individuals. The rigidness contributes to the loss of focusing ability we call presbyopia; the decreased transparency causes a cataract.

There are other less common types of cataracts that are not related to aging. Among these are *secondary cataracts,* which can result from such other factors as trauma to the lens or other parts of the eye, an eye infection, or a metabolic disorder such as diabetes. Secondary cataracts can also be brought about by various forms of radiation, high-voltage electrical shocks, and as a side effect of some drugs. There is also a rare type of cataract that is present at birth and is called a *congenital cataract.*

Depending on the extent and density of the lens disclarity, cataracts can be without symptoms, can have extremely subtle symptoms, or very obvious ones. But in all cases, the symptoms have to do with how well you see, since in no case does a cataract cause pain, excessive tearing, redness, or any other eye discomfort. If the cataract is of the "specks on the window" type, no visual problem will be experienced. Patients with more extensive cataracts invariably come to their doctors saying they need new glasses because things seem less clear to them and the resolution of images is less sharp than it used to be. Very often, however, even a severe cataract can go unnoticed since we all tend to compensate for diminished vision in one eye by relying on the other, stronger eye. This is something we do involuntarily, and we rarely are aware that we are using one eye predominantly. There may be some loss of depth perception since only one eye is delivering a sharp message, but it is usually too slight to be noticed in normal, daily use. Furthermore, cataracts in most cases develop very gradually, so we adjust to the subtle changes in sight over a relatively long period of time.

If You Have a Cataract . . .

When I tell my patients that they have a cataract, they usually ask a lot of questions that indicate their incomplete knowledge of this eye disease. Perhaps you wonder about some of these things too. Let me ask and answer some of the questions I most frequently hear from my patients.

Do you have to change the way you use your eyes? No, you do not. You can read, watch television, attend movies, or do anything you normally do with your eyes as often and as intensely as you like and as is comfortable to you. You cannot make your cataract worse by using your eyes just as you always have, nor can you aggravate a cataract in your other eye or harm it in any way by using it more than your impaired eye. The only change you should make in your way of life is to see your doctor somewhat more frequently than you did before your cataract was diagnosed.

Can you do anything to keep your cataract from getting worse? No, you cannot, but this does not mean that it will necessarily get worse. Many cataracts remain about the same, though others, of course, do worsen, but both courses are out of your and your doctor's control. Nor can your doctor accurately predict whether or not your cataract will remain stable or get worse.

Can you do anything, short of surgery, to make a cataract better? Is surgery always necessary to treat a cataract? There is no treatment for cataracts besides surgery—no drops, exercises, vitamins, or any other therapeutic measures that will either improve your cataract or keep it from getting worse if it is going to do so. But this does not mean that surgery is always necessary. In fact, most people who have cataracts never have surgery. The need for surgery depends on the amount of your visual impairment. If your cataract is small enough and your eyes continue to give you good vision, you do not have to have the cataract removed. A rare exception is if it is causing a secondary medical eye problem, such as glaucoma, in which case your doctor may recommend surgery to relieve this secondary condition. All in all, the decision for surgery is up to you; it is not as though you had a rampant disease that was threatening your total eye health.

Will your sight be seriously impaired after cataract surgery? Will you go blind? The aim of cataract surgery is to improve sight by removing the clouded lens and returning your optical media to a perfectly clear state. There is no need to fear greater visual impairment from surgery. The surgery does make the eye artificially farsighted because the lens is removed, but once your doctor has prescribed corrective lenses, your vision will be vastly superior to what it was before surgery. The fear that removing the lens will cause blindness is quite common, but anyone who understands how the eye works and how we see will realize that this is a totally empty fear. The lens is only one part of the optical system, and its light-bending duties can be taken over by glasses or contact lenses, so it is certainly possible to see without nature's lens.

Cataract Surgery

If your cataract interferes with your sight enough to make you want to improve it, surgery can be scheduled at any time. There are a number of factors that influence the best time to do it, but they are not ironclad. The main one is your own visual needs and how well they are being filled. This is affected in part by the condition of your other eye. A less common factor is a secondary eye problem that might necessitate lens removal. Some people believe that a cataract must ripen and be removed at a specific moment, and that any time before or after that it is not possible to do effective surgery. This is absolutely false.

Although senile cataracts generally occur in both eyes, it is an odd and fortunate fact of nature that one is usually more advanced than the other. This works out well since, before surgery, a person with even two cataracts will usually have considerably better vision out of one eye, and after surgery, the unoperated eye can continue to serve the individual's visual needs while the operated eye is healing. If you have cataracts in both eyes and you and your doctor have decided on surgery, one of the things that will determine when to do it is, therefore, the condition of the cataract in the better eye. If it appears that in the near future that eye will also begin not to meet your visual needs, surgery should be performed on the worse eye while the better eye still functions reasonably well. By the time that cataract needs to be removed, the eye operated on first will be healed and optically corrected and can take over the task of seeing. In this fashion any real period of visual disability is avoided.

It is certainly possible to operate on both eyes at the same time, but it is much less desirable and is rarely done, since the temporary but extreme visual impairment that results when both eyes have undergone surgery at the same time greatly limits an individual's activities. If both cataracts are advanced enough to require immediate surgery, it is usually performed during the same hospital stay, but a few days apart to ensure that no complications have arisen and that the eye operated on first is well on its way to healing.

The surgical procedure itself is not complicated, and the success rate is very high. All the same, the idea of eye surgery of any sort is often quite difficult for people to grasp, and it sounds like a frightening enterprise. Abdominal surgery, though it is often far more serious and debilitating and involves a longer hospital stay and recovery time, inspires far less dread than news of impending eye surgery. For one thing, it is hard to imagine that something as small as the eye can be operated on. For another, the exposed eyes seem so vulnerable, especially if you believe—as many do—the incorrect notion that the eyeball must be taken from its socket for surgery and then replaced. Above all, however, there is the fear of blindness. Although you have probably heard a horror story or two about eye surgery, the operation for cataracts is one of the safest of all surgical procedures. The complication rate is negligible, the success rate is among the highest in medicine, the hospital stay is quite short, the physical discomfort is nothing compared to that experienced with many other sorts of surgical intervention, and the visual restoration, when the eye is otherwise healthy, is remarkable.

The length of the hospital stay depends on your doctor, the procedure he uses, and on you, the patient, but it usually lasts from two to seven days. The operation is performed under local or general anesthesia in the operating room of a hospital. Using tiny instruments and a magnifying glass, your eye doctor makes an incision in the conjunctiva and sclera to get to the iris. He then removes a small portion of the iris, usually in the shape of a tiny slice of pie. This *iridectomy,* which is performed as a routine part of most cataract operations, serves to give your doctor easy access to the lens and to reduce the chance of secondary glaucoma developing in the operated eye. The iridectomy is not visible under normal conditions, but close examination with the lid pulled up would show the missing bit.

Once the iris has been cut, your doctor is ready to remove the lens. There are a number of ways to do this, the most common of which involves freezing the lens, which separates it from surrounding tissues and makes it easier to remove without breaking, a complication much more common with procedures used in the past. After the lens has been removed, tiny stitches are used to close the incision and the eye is patched. There are no evident scars after the eye has healed.

Another procedure is called *phako-emulsification,* in which the lens is dissolved and then suctioned out of the eye. This has a certain appeal because it requires a somewhat shorter hospital stay, but it is not so widely used as the freezing technique and is not feasible for all cataract patients.

Another type combines lens removal with implantation of an artificial lens. I am not alone among members of the medical profession who feel this is a potentially dangerous procedure with added risk of complications. Although the artificial lens is theoretically intended to give patients normal vision without requiring external corrective lenses, it is in practice never possible to determine precisely what the postoperative optics of an eye will be, so a patient with an implanted lens will almost certainly have to wear corrective lenses too. Furthermore, should there be any problem with the implanted lens, a second major operation is required to remove it. With one of the conventional surgical techniques, once your impaired lens has been removed, no further related intervention is required in that eye. Now that certain types of contact lenses are available for around-the-clock wear, even the theoretical attraction of lens implants has become questionable. Surgery is a serious matter, and it seems to me foolish to risk the greater possibility of subsequent surgery for the dubious advantage of an implanted lens.

What Happens in the First Three Months after Surgery

The period of total healing and permanent adjustment in the operated eye may take up to three months. By the time you leave the hospital, the eye patch will probably have been removed and your eye will be open. It may be somewhat uncomfortable and somewhat reddened for a while, and you may notice excessive tearing, but there is generally no pain.

You may be instructed to wear a protective shield for a short period when you sleep to guard your eye, and in some cases your doctor may give you some eye drops.

During the healing period your sight may not be much different from what it was before the operation. Until you get a corrective lens for the operated eye, your unoperated eye will continue to do the seeing as it did before surgery, since it will still be the one delivering the best message to your brain. If you covered that eye and looked out of your operated eye, you would probably see a very blurred picture because lens removal has made you very farsighted. If you were very nearsighted before the operation, lens removal could give you nearly normal vision in the operated eye, since the postoperative farsightedness can, in some cases, balance the optical defect of nearsightedness and bring the point of focus to or quite near the retina. This is rare, however, and most patients will simply continue to use their unoperated eye and see much the way they did before the operation, until they get postoperative corrective lenses.

What Happens Three Months after Surgery

Approximately three months after your lens has been removed, your doctor will determine if the healing is complete and the new optics of your eye have stabilized. If they have, he will measure your optics and prescribe permanent corrective lenses so that the sight in your operated eye will be normal. Many people ask me why vision is not corrected immediately after surgery, either with contact lenses or glasses. Essentially, this can be done, but it customarily is not done when only one eye has been operated on because any correction before the eye has totally stabilized will, of necessity, be temporary. Before the eye heals completely, a contact lens could not be comfortably worn. Glasses do not cause this problem, but the prescription could well have to be changed every couple of weeks. If it is not necessary to make correction because there is no serious visual impairment, most doctors prefer to wait until permanent correction can be made. In the case of a patient who has had both eyes operated on within the same hospital stay or whose unoperated eye fails to provide useful vision, glasses would be prescribed since visual impairment would otherwise be extreme, but these are generally the only cases when the expense and inconvenience of temporary glasses are justified.

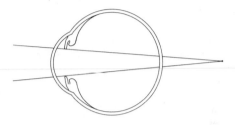

Farsightedness caused by removal of a cataractous lens.

Contact lenses or glasses can be used to correct the extreme farsightedness that results from cataract surgery, but contact lenses are by far the better solution, and if only one eye has been operated on, they are the only effective solution short of no correction at all.

The triple optical system of glasses changes the size of images seen and distorts spatial relations in a way that is disorienting and quite difficult to get used to with the very strong prescription usually required. If both eyes have been operated on, the patient who wears glasses will have to become reaccustomed to the location of objects in relation to each other, a task many people, particularly if they are elderly, find disturbing. For example, in reaching for a cup on a table, the patient with glasses might reach to the right or the left of the handle even though it looked as if the hand were reaching straight for the cup.

If only one eye has been operated on and glasses are worn, extreme spatial disorientation is compounded by the enormous discrepancy between the size of images seen with each eye. This usually makes seeing through glasses intolerable and can cause headaches, dizziness, and nausea. These problems are avoided with contact lenses, since they do not alter the size of images seen or have any unnatural effect on spatial perception. Each eye can be corrected with a lens that suits its optics without causing any visual distortion. Furthermore, contacts avoid the grotesquely large and bulging appearance of the eyes that is a characteristic effect of postoperative glasses. There is no distortion of the eyes when seen by the outside world, so a person who has had cataract surgery will look normal *and* see normally with contact lenses.

If a patient simply cannot tolerate a contact lens, and only one eye has been operated on, the only solution is to correct the eye with the greatest potential for normal vision and leave the other uncorrected. The corrected eye will dominate and vision will be good, but depth perception will be limited.

Whenever possible, however, I try to encourage my postoperative cataract patients to give contact lenses a try. Most have no difficulty at all and are able to avoid the readjustment to spatial relations that those who choose glasses must undergo. Patients who cannot or prefer not to remove their lenses before going to sleep can be fitted with a type of soft lens intended essentially for full-time wear. I remove and clean and then reinsert the lenses during these patients' periodic check-ups, and for the rest of the time they can see well and be virtually unaware that they have anything in their eyes.

One problem postoperative patients tend to have is a greater sensitivity to light, since a portion of their iris has been removed and the slight light-filtering properties of the lens are no longer available to them. This can be solved with tinted glasses or contact lenses, and with sunglasses when the patient is outside. Other than that, once corrective lenses have been prescribed and adjustment to them made, a postoperative patient will have optically corrected and cataract-free eyes. The cataract obviously cannot "grow" back, since the lens is no longer there.

Although senile cataracts lie in the future for the majority of us, I hope it is now clear that they are not a dread disease that inevitably threatens sight and guarantees a date in the operating room. By far the greatest number of people who have cataracts never have them surgically removed and are able to see adequately in spite of the imperfectly clear lens. Those who do have surgery do so voluntarily and make their decision in partnership with their eye doctor. Because cataracts rarely endanger the health of the eye as a whole, the decision to treat them surgically is based on personal visual needs and how well they are being met. I try to help my patients make an intelligent decision by letting them know what sort of improvement they can expect if they choose surgery and when I think it is desirable to avoid a period of serious visual impairment, but the choice is ultimately their own. And because cataract surgery has an extremely low risk factor and the prospects for optical correction after surgery are very rosy indeed, no one need fear blindness as the result of a cataract.

9

Glaucoma

MYTH: You will be able to tell if you have glaucoma because you will experience eye pain, see halos around lights, have excessive tearing, or your eyes will bother you in some other way.

FACT: Although there is a rare form of glaucoma that can be quite painful, the most common type causes no pain at all and is without symptoms until the disease is far advanced.

MYTH: Glaucoma is a cancer, a tumor, or an infection.

FACT: Glaucoma is none of these. Rather, it is a disease of the eye in which pressure inside the eye is higher than it should be and, if it goes untreated, gradual but permanent damage to sight can result.

MYTH: You should not drink liquids, especially coffee and alcohol, if you have glaucoma.

FACT: This is a holdover from years ago when glaucoma patients were advised to reduce their intake of liquids. Modern methods of treating glaucoma make this unnecessary. A glaucoma patient today can lead a perfectly normal life as long as prescribed medication is taken as a doctor directs.

MYTH: High blood pressure and the high eye pressure of glaucoma are somehow related.

FACT: There is no connection whatever between elevated blood pressure and the increased eye pressure that causes glaucoma. People with high blood pressure are not more likely to develop glaucoma than people whose blood pressure is normal, nor are glaucoma patients more prone to high blood pressure.

Of all the possible causes of blindness, glaucoma is the most common, but it is also the easiest to prevent. If undiagnosed and untreated, it will gradually cause the loss of sight; if diagnosed, glaucoma can be controlled and the loss of sight prevented. It is really as simple as that. What complicates matters is that in its most common form, glaucoma has no symptoms until extensive and irreversible damage has been done, so that the only way it can be diagnosed early is by an eye doctor. Because glaucoma generally occurs after middle age, people over forty should have their eyes examined more frequently than younger people. I recommend once-a-year eye checkups after forty, whether or not you wear glasses or contact lenses.

It is relatively simple for an eye doctor to spot glaucoma, and the rate of successful treatment is high. All the same, each year glaucoma is responsible for countless cases of blindness or extensive, permanent loss of vision. In the overwhelming majority of these cases, the damage could have been avoided. If all persons over forty visited their eye doctor annually, glaucoma-induced blindness could be virtually eliminated.

What Is Glaucoma?

Glaucoma is an eye disease characterized by a higher than normal pressure inside the eye *(intraocular pressure)*. Pressure can increase for various reasons, and it can happen gradually or suddenly, slightly or dramatically, depending on the type of glaucoma. By far the most common type is called chronic *simple glaucoma* (see also *chronic glaucoma,* page 193), in which the pressure increase is small, the beginning stages of the disease without symptoms, and the progressive damage slow and gradual. It occurs most commonly after age forty; in fact, about one out of twenty-five people past that age have chronic simple glaucoma. Because it is the most common type, I will focus my discussion on chronic simple glaucoma, and touch more briefly on rarer forms of the disease.

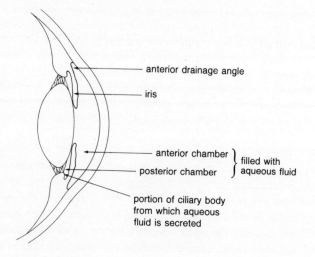

Where the aqueous fluid enters and drains from the eyeball.

Let's look at what happens inside the eye when intraocular pressure increases. You know that the inside surface of the cornea is nourished by the aqueous fluid, which is secreted by the ciliary body into the area behind the iris and flows through the pupil into the area in front of the iris. The aqueous fluid is reabsorbed and carried away to the blood at a point called the *anterior drainage angle,* located where the front of the iris joins the back of the cornea. It is at this location that problems arise. In a healthy eye the drainage network is in efficient working order and the balance between secretion and drainage of aqueous fluid is maintained by a relatively constant intraocular pressure.

With glaucoma, something goes wrong with the drainage part of the process. For one of a variety of reasons, some of which are not well understood, the drainage mechanism is less efficient, the secretion/drainage balance is upset, and intraocular pressure is increased. This does not mean that there is too much aqueous fluid in the eye, since the pressure increase forces the fluid to drain. It can happen in one eye or both, though it is most common for it to affect both eyes.

Increased intraocular pressure, even when it is very slight, affects the eye. The treatment for glaucoma is not designed to correct the underlying problem in the drainage mechanism; it aims instead to bring the pressure back down to normal. In this sense, glaucoma cannot be cured, but it can be controlled so that damage to the eye is avoided.

What happens when intraocular pressure is too high? We know that the eyeball has a very effective system of protective layers to prevent injury to the delicate internal structures of the eye. The sclera, which covers the entire eyeball except for the cornea and the place where the optic nerve enters the eye, is very tough, as is the cornea, but the optic nerve, the vital connection between eye and brain, is not so tough, and it is vulnerable to elevated pressure. The increasing pressure affects the entire eye, making the eyeball harder than it should be and causing stress to all parts. The eye as a whole can stand this without being damaged, but the delicate nerve fibers and blood vessels at the end of the optic nerve (the *optic disc*) cannot. If the pressure continues long enough, it will kill nerve fibers in the optic nerve, and once they have died they cannot be revived.

We refer to glaucoma as a progressive disease, but it is really the damage it inflicts that is progressive. The defect in drainage does not necessarily get worse, the pressure does not tend to increase steadily, but if untreated, the effect—the killing of the cells of the optic nerve—continues and the resulting loss of vision becomes progressively greater until total death of the optic nerve results in total blindness.

The nerve fibers at the outer edge of the optic disc are the first to go. In time, cells farther in toward the center die off. The outer edge corresponds to the outermost periphery of the retina and therefore to our outermost peripheral vision. Each succeeding layer of nerve fiber corresponds to a more central area of the retina, and the innermost layers contain the nerve fibers that receive messages from the macula. As each layer dies, the peripheral visual field narrows, and finally central vision is lost. In the later stages of untreated glaucoma a person could have very narrow tunnel vision and be able to see nothing at all off to the sides, but still have 20/20 vision when looking at objects straight ahead, since it is the macula, and the macula only, that can deliver a 20/20 image to the brain.

Uncontrolled glaucoma can result in progressive narrowing of the field of vision: *(left)* controlled glaucoma, with no visual loss; *(center)* uncontrolled glaucoma, with some loss of peripheral vision; *(right)* advanced uncontrolled glaucoma, with serious loss of peripheral vision.

People with untreated chronic glaucoma are rarely aware of the disease until a great deal of damage has been done, since the progressive death of nerve fibers takes place very slowly and the elevation of pressure is usually too slight to cause pain or blurring of vision. Besides, we are all mostly unaware of what we see on the edges of our visual fields, since the messages received from that area of the retina are not very sharp and provide little more than our sense of the surrounding landscape. But even if the gradual loss of peripheral vision is imperceptible to us, it can be spotted by an eye doctor when he measures and diagrams the visual field.

Types of Glaucoma

Chronic simple glaucoma is by far the most common type. The word *simple* in this case means that the pressure rise is not the result of any known underlying factor, for, in fact, we do not know exactly what causes it to happen. Heredity seems to play some part but not a major one. This is why your doctor will test for glaucoma even before you reach forty if there is a history of it in your family, but it certainly does not mean that everyone with a family history will develop the disease, or that it is impossible to find glaucoma in a person whose family has no history of it.

The fact that chronic simple glaucoma often begins in the middle years does not explain the cause either. Glaucoma is not, in this sense, like senile cataracts, which are part of the aging process and result from eye changes that always take place as we grow older. The closest we can come to an explanation is to say that glaucoma seems to develop in people who have a tendency toward inefficient drainage of aqueous fluid, and as these people grow older and their bodies as a whole begin to lose their resiliency, the drainage efficiency lessens.

Chronic *secondary glaucoma* has all the distinguishing features of chronic simple glaucoma except that the drainage defect is caused by something identifiable, most commonly a complication of another eye problem. Inflammation from an internal eye infection can be responsible for poor drainage; likewise, an allergic reaction within the eye, trauma to the eye, or scar tissue left by any of the above can reduce drainage efficiency. Chronic secondary glaucoma can also be a complication of cataracts, since the diseased lens, which tends to be enlarged, can encroach upon the front of the eyeball and narrow or block the drainage angle.

Acute glaucoma, whether simple or secondary, is a rare and dramatic condition. The elevation of intraocular pressure is rapid and many times higher than ever experienced with chronic glaucoma. The drainage angle is almost totally blocked, which causes the eyeball to become very hard. The pressure increase in chronic glaucoma is so slight that it can be detected only with special instruments, but in acute glaucoma the increase can often be felt by touching the front of the eye with your fingers. The pressure is so great that it causes severe pain and damage to the whole eye; the cornea can become clouded, which results in blurred vision; blood vessels in the eyeball become swollen, causing pronounced redness; and the nerves surrounding the blood vessels respond to the pressure, causing great pain. Headaches, nausea, vomiting, and abdominal pains are frequent symptoms of an acute glaucoma attack. The destructive effects on the optic nerve are equally rapid and severe, and immediate treatment and surgery are required.

Acute glaucoma can result when a drainage angle that has always been abnormally narrow suddenly becomes completely blocked; in this case it is called *acute simple glaucoma*. Or it can be a complication of a number of eye problems, including infection, allergic reaction, trauma, or cataracts. It is then called *acute secondary glaucoma*.

A third type of glaucoma, also quite rare, is called *congenital glaucoma*. It is present at birth or develops during infancy, and it is the result of an anatomical defect in which the drainage angle is very narrow or the drainage canals are malformed. It most commonly occurs in both eyes. The pressure elevation tends to be higher than in chronic glaucoma but not so extreme as in acute. Its characteristics differ from both types, mostly because a baby's eyeball is much less tough than an adult's. Surgery is usually necessary, and it must be performed early if permanent loss of sight is to be avoided.

A very rare fourth type is called *low tension glaucoma*. In this instance, an eye exhibits signs of glaucoma damage even though the pressure is within the normal range. The explanation is that this particular eye requires a lower than normal eye pressure to avoid damage to the optic nerve, so the pressure level normal for others is exerting stresses in that eye. But be careful not to confuse low tension glaucoma with low eye pressure. A severe perforating injury to the eye may cause aqueous and, possibly, vitreous fluid to drain from the eye, which will bring about severely low eye pressure. This is analogous to low blood pressure when a person is in shock, and it is an emergency situation. But this is the only time when low eye pressure is a medical problem. And, of course, intraocular pressure has nothing to do with blood pressure.

How Is Glaucoma Diagnosed?

Glaucoma cannot be prevented or cured, but if it is diagnosed early, it is generally easy to control so that no further damage is caused. Any damage already done cannot be repaired, so the sooner the disease is diagnosed and treatment begun, the sooner the effects will be curbed.

If a person has glaucoma, I can spot several signs in the course of a routine examination. In the first part of the exam I perform on all patients, regardless of age or family history of glaucoma, I examine the inside of the eye. At this time, I look at the relative size and depth of the anterior chamber. If it looks shallow or abnormal in any way, I look sideways at the drainage angle to see if it is blocked or narrow or normal. Later in the routine exam, I put drops in the patient's eyes, which dilate the pupils so I can look at the inside back of the eye, including the optic disc. Certain abnormalities in its shape or color are clues that glaucoma may be present and nerve cells are dying.

I also measure the intraocular pressure of all patients over forty and of younger patients who have a family history of glaucoma or other eye problems that can be secondary causes.

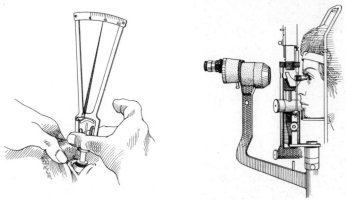

Two ways to test for glaucoma.

Pressure is measured with an instrument called a *tonometer*. There are basically two ways to do the measurement, but both tell your doctor what your intraocular pressure is, and in both cases your eyeball is anesthetized with drops so the procedure is completely painless.

If the pressure reading as well as the other parts of the routine exam pertinent to glaucoma are within the normal range, I can conclude that my patient does not have glaucoma and that it is safe to let a year elapse before testing again. If the pressure is above or at the upper limits of normal, I have to do more sophisticated testing. First I test the peripheral visual fields, since any narrowing of the field is possible evidence of glaucoma damage.

There are two main types of instruments used for this test, both of which allow your doctor to obtain a diagram of the shape and size of your visual field, as well as the shape, size, and location of your blind spot. One instrument uses a large blackboard-like screen hanging on the wall. You sit in front of it with one eye covered, and stare with your other eye at a white dot in the center of the screen. As you stare, your

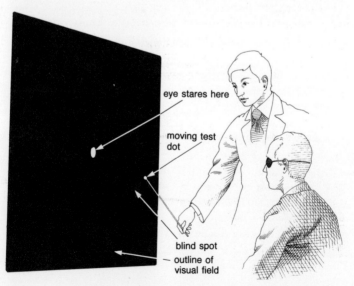

eye stares here

moving test dot

blind spot
— outline of visual field

Testing visual fields with the tangent screen.

doctor will bring a second white dot (mounted on the end of a black stick, which is invisible against the black background of the screen) into view from all the various peripheral areas. You will be asked to say "yes" as soon as you see the second dot. Each eye is tested in turn. The other type of instrument records your responses and produces a diagram of your visual field electronically. You rest your chin on a platform facing

Testing visual fields with the perimeter.

into a machine so that all extraneous visual stimuli are excluded, and one eye at a time is tested. Small blips of light rather than the white-tipped stick are set off on your periphery, and again you indicate when the blips become visible to you. Regardless of which instrument he uses, your doctor ends up with a picture of your peripheral visual field and blind spot for each eye, so any abnormality in size and shape will be obvious.

But a positive diagnosis of glaucoma can be made even if the disease has not progressed to the point of damaging vision. If the drainage angle looks normal and there is no narrowing of the visual field, but I noticed a borderline pressure elevation, I can refine my diagnosis with provocative testing.

Intraocular pressure does not always remain the same. There is a normal range within which the pressure can vary somewhat at any given moment. For example, it is normal for pressure to be somewhat higher in the morning and when the pupil is dilated, since this mechanically narrows the drainage angle. A reading in the upper range of normal suggests to me that pressure may sometimes be even higher. What the provocative test tries to do is aggravate the situation temporarily so that if there is a tendency toward poor drainage it will show up.

The most commonly used provocative test has the patient drink a large quantity of water and sit in a pitch-dark room. The excess liquid intake increases the secretion of aqueous, and the darkness makes the pupils dilate. In combination, these factors make drainage more stressful than normal. I take pressure readings before and after applying the stress. If the patient's eyes can handle drainage under these conditions with a minimal increase in pressure, I can say that there is no glaucoma. If I observe a significant increase in pressure, it is clear there is something wrong with the drainage mechanism.

Sometimes another special test is helpful in diagnosing glaucoma. Called *tonography,* it electronically measures the efficiency of the drainage mechanism.

There are three possible conclusions that can be drawn from a complete glaucoma evaluation. If all tests are negative, the patient does not have glaucoma and need not be tested again for another year. If the tests are positive, the patient has frank glaucoma and must be treated. If some tests indicate a tendency toward poor drainage, this is what we call a *glaucoma-suspect* situation and the patient must be more carefully watched. This patient may never develop frank glaucoma, but it is still essential that people in this category be checked more frequently than those who show no evidence of drainage inefficiency. The patient with diagnosed glaucoma must, of course, be seen by an eye doctor even more often. I see my glaucoma patients every three months, and my glaucoma-suspect patients once every six months.

Treatment of Glaucoma

There are various ways of treating glaucoma, but it is a general medical principle that the simplest treatment that does the job is the one to choose. In the case of glaucoma, the job is to lower the intraocular pressure and keep it within the normal range. How this is done differs between chronic and acute glaucoma, and between simple and secondary glaucoma of both types.

Constricted pupil with enlarged drainage angle.

Dilated pupil with narrowed drainage angle.

For chronic simple glaucoma, the simplest treatment is to use special drops that constrict the pupil, which in turn mechanically enlarges the drainage angle to make drainage more efficient. Some types of drops also inhibit secretion of aqueous fluid so there is less to be drained. I always prescribe the drop that is weakest and can be used least frequently to keep the individual patient's pressure normal.

These drops are safe and convenient to use. They are generally free of side effects and, because they are administered locally, they do not affect the body as a whole. When first used, they may cause a slight pulling sensation around the brow area and a slight blurring of vision, but these effects disappear within a few days. Drops will also keep the pupils from dilating as much as before, but this does not interfere with vision in any practical way. Patients who use drops must do so regularly for the rest of their lives, but if they do, their glaucoma will be under control, the nerve cells will not be threatened, and if their eyes are otherwise healthy, they will experience no further eye problems.

If the highest dosage of the strongest drops cannot reduce pressure and maintain it within normal limits, I next try drops in combination with tablets that work to decrease aqueous secretion rather than affect the drainage mechanism. The tablets are less desirable than drops because they are a systemic rather than local medication and can, therefore, affect the body as a whole, but they certainly are not dangerous. And, again, patients undergoing this treatment must continue the regimen for the rest of their lives.

People whose glaucoma is under control can lead normal lives. They do not have to avoid drinking liquids or sitting in the dark, since controlled glaucoma will not be aggravated in this way. Nor is it, incidentally, possible to give yourself glaucoma by increasing drainage stress. The eyes of a person with no drainage defect will be able to handle increased secretion of aqueous and/or slight narrowing of the drainage angle when the pupil is dilated without a significant rise in intraocular pressure.

Except in the case of acute or congenital glaucoma, medical management with drops and/or pills is always preferable to surgery, the last resort if all other therapy fails. In the past decade new and stronger medication has made it possible to manage successfully most cases of glaucoma, so surgery has become an infrequent treatment.

There are a variety of surgical procedures in common use, but they all try to do the same thing—keep intraocular pressure within normal bounds. One type of procedure does this by creating new or improved drainage outlets for the aqueous. Another enlarges the already existing drainage angle by removing a portion of the iris. (This is the iridectomy that is routinely performed as part of cataract surgery, partially to guard against secondary glaucoma.) A less common approach uses a tiny ice probe on the sclera to freeze the ciliary body, which results in decreased aqueous secretion. Because this also affects the ciliary muscles, it is usually performed only on older patients who have already lost their near-focusing ability or on those who have had cataract surgery and, therefore, have no need whatever for the ciliary muscles.

Sometimes drops or tablets are required to lower pressure

further if the surgery has not succeeded in lowering it enough. If surgery—combined with the strongest possible medication taken as many times a day as can be recommended—still does not control pressure, a second operation may be necessary.

The length of the hospital stay for glaucoma surgery varies, but it is usually just a few days. A patch, if used at all, rarely needs to be worn for more than two or three days after the operation, and most people experience only slight discomfort as the eye heals. Some slight change in the eyes' optics may result from the surgery since the interior of the eye has been operated on. As soon as the eye has healed and the new optics have stabilized, I recheck my patient's eyes and prescribe a new correction, if necessary. There are no visible scars on the healed eye, nor does it appear in any other way different from normal.

If the chronic glaucoma is secondary, I treat it in one of the ways outlined above, but, of course, I also treat the underlying problem whenever possible. If the secondary glaucoma is caused by a developing cataract, for example, removing the lens might control the glaucoma at the same time as it eliminates the cataract.

The treatment for acute glaucoma, whether simple or secondary, is immediate surgery. The pain, the near-total blockage of the drainage angle, and the resulting extreme high pressure make this an emergency situation, but even when surgery is performed right away, some damage to the optic disc and some loss of peripheral vision is usually unavoidable. If the acute glaucoma is secondary, the underlying cause must also be treated if at all possible.

Once diagnosed, glaucoma can be treated with great success. Treatment generally involves an easy-to-follow regime that must become a daily habit for the rest of the patient's life. Regular visits to the eye doctor are an essential part of the treatment.

Approximately 4 percent of the over-forty population in the United States develops glaucoma, and every year large numbers suffer severe damage to sight or total blindness as a result. This is tragic since accurate and early diagnosis is a relatively simple matter and control of the disease is in most cases complete. Making yourself available for periodic preventive eye examinations is all you have to do to protect yourself from the threat of glaucoma-induced blindness. This is a small enough commitment to make in exchange for healthy eyes and good vision.

10

A Handbook of First-Aid and Common-Sense Eye Care

MYTH: Sitting too close to a television set or movie screen is bad for your eyes.
FACT: You cannot injure your eyes in any way by sitting very close to a TV set or in the front row of a movie theater. You should sit wherever you feel most comfortable.

MYTH: Reading in bad light can ruin your eyes.
FACT: The quality and level of light in which you read cannot have any effect on your eyesight or eye health. It is easier and more comfortable to read in good light, but, again, comfort, not eye health, is the issue.

MYTH: You should wash your eyes regularly with an eye wash.
FACT: Your natural tears are the best and only eye wash necessary. They lubricate your eyes and wash away pollutants every time you blink.

MYTH: There is a proper level of room light for watching television, and if you watch in the wrong light you will hurt your eyes.
FACT: This is untrue. The "proper" light level is what is comfortable for you. Some people feel there is more contrast

if the room in which they watch television is dark. If you are one of these people, by all means turn off the lights. You will not be hurting your eyes, but neither will those who watch television in a lighted room.

MYTH: Fluorescent lighting is bad for your eyes.
FACT: There are no health dangers or benefits to be derived from either fluorescent or incandescent light. Many people find fluorescent lighting harsh and therefore avoid it when possible, but you will not be endangering your eyes with it.

We all tend to take our health for granted when it is good, and the health of our eyes is no exception. But when things go wrong with our eyes, we worry more than when things go wrong elsewhere in our bodies. None of us takes a mild stomach-ache seriously, but we often over-react to an equally mild eye ache. This need not be so. Although there are certainly serious eye problems and diseases, those most commonly experienced are minor and easily treated, if indeed they require treatment at all. And by far the most common experience we have with our eyes is that we have no problem at all.

Let's turn our attention now to some common eye problems and discuss which are minor and which are emergencies, which ones require medical treatment and which you can treat yourself. I will also give some advice about what you can do to avoid some of these problems so that your healthy eyes will continue to serve you well.

What Is a True Eye Emergency?

Because the human eye is pretty tough and well protected, there are only four true eye emergencies, cases when sight is endangered and immediate medical care is required. An acute glaucoma attack is one of these. The symptoms are severe pain in the eye, pronounced redness of the eye, blurred vision, and possibly abdominal pain and nausea. Do not try to diagnose acute glaucoma yourself. If you experience such symptoms, go immediately to your eye doctor or to a hospital emergency room. Acute glaucoma is quite rare, but it should be attended to without delay to guard against the danger of severe visual loss.

The second sight-endangering emergency is when a high-velocity foreign body enters the eye and pierces its protective layers. I am not talking about particles of grit or other airborne bodies that might land on the conjunctiva or cornea, nor a finger, pencil, or other object that has been poked into the eye. Although these can be painful, they are not likely to damage your sight. What I mean here is a sharp, hard object traveling at a speed fast enough to pierce through the protective layers and enter the interior of the eyeball. This sort of injury usually occurs when a piece of metal has been thrown off from high-speed power equipment in a factory or a workshop.

This kind of injury may take place without causing much pain. If the object is tiny and traveling fast enough, the place where it enters the eye may seal up immediately, so that little or no pain will be felt. Nonetheless, it can have catastrophic results and is to be avoided at all costs. If you work with power tools or other high-speed equipment, you must wear protective goggles. I advise patients who have good vision in only one eye to avoid such situations altogether. If such an injury occurs, you must get to an eye doctor or a hospital emergency room as soon as possible. I do not recommend any do-it-yourself first-aid measures, since the time required to administer first aid can often be a waste, and there is no first-aid treatment of value.

There are two other eye emergencies, but for these you can and should do something before rushing to get medical attention. Speed and calm are essential in these cases, but the first-aid measures you take may be able to prevent serious injury to the eye.

The first involves severe injury in or around the eye, as when flying glass or jagged metal has cut the face. (An automobile accident is a common instance of this.) While you are taking steps to get emergency medical treatment, you should try to stop the flow of blood by applying pressure to the wounded area with a clean cloth. Do not try to do this if applying pressure will drive any glass or metal farther into the eye or skin.

The other remaining eye emergency occurs when a very strong chemical, such as lye, gets in your eye. It is absolutely essential that the eye be flushed with large quantities of water as soon as possible. Do not wait for a doctor to do this; you can do it yourself (I will tell you the best way a little later, see p. 151). The time saved can truly be sight-saving.

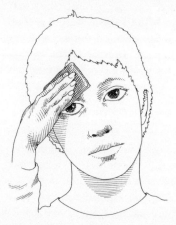

Apply pressure to stop bleeding from a cut around the eye.

Some First-Aid Measures You Can Take

Aside from the four emergency situations mentioned above, injury to the eye can be treated at home or given first aid before you go to your eye doctor.

Perhaps the most common non-emergency eye problem people experience is getting something in their eye. Depending on what and where it is, something in the eye can be either a minor irritant or very painful. The first thing to do when something gets in your eye is to remain calm and to proceed as quickly as possible to a well-lighted mirror. In the meantime, your eye will be trying to solve the problem on its own

by shutting involuntarily and secreting more than the normal amount of tears. Both of these reflexes are beneficial: shutting the lids will avoid scratching the surface of your eyeball, and the tears will tend to wash the particle away. In many cases you may find that whatever was in your eye is gone by the time you arrive at the mirror.

If the particle is still there, do not rub your eye or try to remove it with your fingers. If you can see the particle in the

You can try to remove a foreign body from your eye with a clean handkerchief. Don't try it if it's on your cornea.

mirror, you can try to remove it with the corner of a clean handkerchief or some other lint-free clean cloth. You may find it easier to have someone do this for you, but it is certainly possible to do the job yourself. If you do not have a handkerchief, do not try to remove the particle with a tissue or cotton-tipped swab; these will probably introduce additional smaller foreign bodies. And do not try to remove it in this way unless it is on the white of your eye or on the inner eyelid. The danger of abrasion is simply too great when a particle is on the cornea.

If you cannot see the particle, if it is on the cornea, if you do not have a clean handkerchief, or if you are simply uneasy about poking around in your eye, you can try to float it off. Fill a sink, bowl, or eye cup with lukewarm tap water and open your eye underwater. If you do not have access to any of these, you can cup your hand and fill it with water. If it does not seem to have worked the first time, try again, moving your eyeball around underwater as much as possible. Although you can use an eye wash for this, it is not necessary and will not be more effective than tap water. If you still feel the particle in your eye, try pulling your upper eyelid over your lower lid. Hold on by the lashes and then quickly release the lid to dislodge the particle. You can try this two or three times.

You can try to float a foreign body out of your eye by keeping the eye open under water.

The particle and all this poking around your eye will tend to cause irritation, and in many cases the particle is long gone even though it still feels as if something were there. Wait a few minutes to see if the irritated feeling goes away. Close your eyes and relax a bit, then blink a few times, letting your tears flow around your eyeballs. Close your eyes again. In most cases, the irritated feeling will subside.

Although you may feel the particle in a particular place in your eye—under the upper lid on the right side, for example —it may in fact be somewhere else. It is hard to localize a foreign particle accurately. Keep this in mind when you are looking for it and also if your doctor is trying to remove it.

I have often heard patients insist that I was looking in the wrong part of their eye since I *saw* it in a different place from where they *felt* it.

If you still feel the particle and all these attempts have failed to remove it, go to your eye doctor or the emergency room of a hospital.

A particular problem with young children as well as adults occurs when they get a liquid or solid chemical substance in

Sometimes you can dislodge a particle by pulling down and then releasing your upper eyelid.

the eye. The substance can be a minor irritant or a major catastrophe, as with lye or strong acids. Even if you don't know what the substance is, the first thing to do is to flush the eye quickly and thoroughly with clear water. Do not try to figure out what the substance is and devise an antidote. This can use up precious time. Flushing with lots of water will dilute and wash the substance from the eye.

The fastest and most effective way to deal with this situation is to rush to the nearest sink, cup your hands, and fill them with water. Then hold your hands to your eye and try to open the eye underwater. Meanwhile, permit the sink to fill up so you can put your eyes underwater, leaving your hands free to force your lids open. Continue the flushing for several minutes after any pain has subsided. If pain persists, if your eye is red, if you experience any loss of vision, or if you know or suspect that the chemical is very strong, grab the container and rush to your eye doctor or hospital emergency room.

If a child has gotten a liquid in his or her eye, you can perform the flushing. Remain calm and try to keep the child quiet so that a thorough flushing of the eye can be done as quickly as possible.

The best way to avoid such an emergency, of course, is to keep household and other chemical substances away from children and to use them carefully yourself. Take particular care not to put your hands in or near your eyes when you are working with them, and be especially careful with aerosol cans. No matter what their other ingredients are, aerosol cans contain propellants that are harmful to the eyes. Hair sprays and spray deodorants commonly find their way into people's eyes. You should always close your eyes when you are using these products and keep your eyes closed for a few seconds after you have finished, since the vapors hang in the air. If it is not possible for you to keep your eyes closed (in the case of spray paints, for example), shield your eyes, keep the can at arm's length, or wear goggles of some sort. And by all means, look to see in what direction the nozzle is pointing before you depress the button!

If you burn your eyeball from splashing hot oil or hot water, the first step is to flush your eyes. The fact that oil is not soluble in water is irrelevant, since you are flooding it out, not trying to dissolve it. Cool water will also tend to reduce the temperature of the burned tissues. If the burn is serious and painful, you should see your doctor. Do not, in any circumstances, try to apply burn ointments, butter, or anything else intended for skin burns.

Injury to the cornea, regardless of the cause, is more serious than injury to the conjunctiva. This is why you should not attempt to remove a particle from the cornea. If you get something on your cornea and cannot flush or blink it away, have a doctor remove it. And if your cornea becomes scratched or abraded, this too will require medical attention. What you can do in the meantime is to try to keep your eye closed, since the main discomfort is caused by blinking. If you wish, you can devise your own eye patch with sterile gauze and adhesive tape.

A corneal burn is another painful injury that must be treated by a doctor. There are a number of ways your cornea can become burned, but the burns I see most often are from sunlamps. The light of the sun is considerably less intense than the artificial sunlight provided by ultraviolet sun lamps. You cannot, therefore, expect sunglasses to protect your eyes adequately from sunlamp light. When you use a sunlamp, be sure always to wear the special goggles provided *and* to keep your eyes closed. This should be done even if your back is to the lamp, since the reflected light is sufficient to cause a burn. Do not lie down and read with your back to the lamp and expect to be safe from the danger. The goggles must be worn, and your eyes must be closed.

You can, of course, also get a burn on your eyelids from a sunlamp or from the sun itself. This is no different from a skin burn elsewhere on your body, although it may feel more painful since the skin of your lids is quite tender. In most cases, the irritation will go away in a day or two. You can, if you like, apply a commercial sunburn cream or lotion, but do it carefully, making sure to get nothing in your eyes. Many of these preparations contain mild local anesthetics to relieve discomfort, but they also often contain chemicals that can sting the eyes. You might also find that a cold compress (a clean washcloth dampened with cold tap water placed over your closed lids) or an ice pack will reduce swelling from the burn and cool your skin.

Minor cuts near the eyes should be treated like cuts elsewhere on the body. If there is bleeding, apply pressure to the area until the bleeding stops. Use sterile gauze, if you have it around, or a clean handkerchief. If the cut is large or the bleeding continues, seek medical attention. Apply antiseptic and a Band-Aid to small cuts and abrasions, as you would with any cut, to keep the wound clean and protected from additional injury. But take care to keep antiseptic out of the eye.

A black eye is really nothing more than a bruise, showing up bluish-purple at first and then gradually fading in from ten days to two weeks. There is nothing you can do to make it fade faster, and that includes using steak on the area. I am not sure where this remedy first originated, but in these days of high meat prices all I can say is buy a good steak so you can eat it afterward. It will not help your eye, but a juicy steak dinner might cheer you up. You can reduce the swelling somewhat with an ice pack, and you will not do any harm to your eye if you cover the bruise with flesh-colored make-up.

Aspirin can be an effective pain reliever in any of these instances. If you can take aspirin or one of the aspirin substitutes, by all means do so, two tablets every four hours (for adults), for as long as you need mild pain relief. The recommended dosage of a children's aspirin can alleviate minor pain for your child.

Eye Drops and Eye Washes

Short of treating eye injuries, isn't there something you should be doing to maintain the health and comfort of your eyes? Other than getting your regular checkups, there really is not. This may come as a surprise to many of you who feel that eye drops, eye washes, or vitamin supplements must be helpful to the eyes in some way, but the fact is that nature has provided our eyes with their own protective and maintenance systems, and no additional help is necessary.

Our tears and blinking reflex lubricate and clean our eyes constantly. You do not, therefore, need eye washes or eye drops to do this job. Commercial eye washes and drops can be beneficial if you happen to hold stock in a pharmaceutical company that manufactures them. The rest of us should not waste our time and money on their use.

What about red eyes? Many people are concerned about this condition, particularly since we are exposed to a great deal of advertising that implies that red eyes are a common and significant problem. The fact is that we all have small blood vessels in the membrane that covers the white of our eyes. In some people they are more pronounced than in others, making their eyes look redder. This is an anatomical variation, not an indication of any sort of disorder or eye-health problem. Your eyes can also become redder than is normal for you as a result of smoking, late hours, air pollutants, drinking, or long, strenuous use. But despite some discomfort and an appearance you might not find pleasing, there is nothing basically wrong with your eyes.

There are no over-the-counter medicines that you need ever buy for red eyes, so don't walk into your local drugstore and ask what they have that's good for red eyes, because they won't have anything. Either you do not need any medication at all for your eyes or you have a medical eye problem, in which case you need your doctor's advice and a prescription from him.

If you feel you want to use eye drops or an eye wash for a soothing effect, the best are those referred to as *artificial tears* —thickened synthetic versions of our natural tears. They are nonmedicinal, available without prescription, and completely harmless. They are not the same as the popular patent eye preparations that claim to reduce redness. These drops contain very weak decongestants, which constrict your eye blood vessels slightly for a few minutes, long enough for you to inspect your eyes in the mirror. A short while after that, all effect has vanished. I feel that these drops are worse than useless—they are a waste of money.

A popular home remedy for tired, red eyes is a compress made from wet tea bags or slices of cucumber. This may appeal to those of you who like "natural" approaches to health, but they are no more or less effective than anything else. You will certainly not be injuring your eyes with tea bags or cucumbers, but they will not benefit your eyes either. If you like the idea, go ahead and follow it; it is cheaper than over-the-counter eye drops, and if you happen to have a black eye too, you can eat a cucumber salad with your steak and wash it all down with a cup of hot tea.

Certainly the most economical way to soothe your eyes is to use plain water. Applying a cool compress or even rinsing your eyes with cool water from the faucet will do as much as eye preparations you can buy in a drugstore. Water cannot harm your eyes, and if you find it effective for temporary relief of itchy or irritated eyes, you should feel free to use it.

The notion that taking vitamins, particularly vitamin A, will somehow help your eyes is a foolish one. If you eat even a reasonably adequate diet, you are getting all the vitamins your eyes can possibly use. Do not waste your money on vitamins for your eyes.

How to Apply Eye Medication

Many people find eye drops and ointments that their doctor has prescribed much more difficult and frightening to administer than pills or ointments designed for other parts of the body. Although you may prefer to have someone else administer eye medicines, it is really quite simple to do it by yourself. Here is the easiest method:

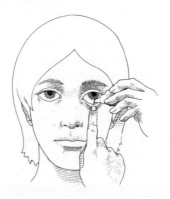

The proper way to put in eye drops.

Stand in front of a mirror in good light. Do not try to put in drops standing over a mirror that lies flat on a table. And don't try to administer drops lying down. This "bombs away!" procedure rarely results in a direct hit. Work on one eye at a time, pulling down the lower eyelid with the fleshy pad of your index or middle finger, not with the tip or nail. Bring the dropper or dropper bottle to your eye with your other hand

and squeeze gently, so a drop lands in the cul-de-sac you have exposed. Be careful not to touch the dropper tip to your eye. Release the lid. Your tears and blinking mechanism will wash the medication over your entire eyeball.

If you are not sure you got the drop into your eye, put in a second one. Any excess will run out, so you are not harming your eye by putting in more. But even if you made a direct hit, some of the liquid will run out of your eye. This does not mean you did not get any or enough medication into your eye.

Use the same procedure for applying eye ointments. In this case, simply squeeze about a quarter of an inch of the medication into the bottom of the cul-de-sac and close your eyes. And again, keep the tip of the tube out of contact with your eye.

Eye ointments leave a slight film over the eye that interferes somewhat with seeing. Their advantage is that they work longer than drops do. For this reason, when necessary, I usually prescribe medication in ointment form for nighttime use and in eye-drop form for use during the day.

If your doctor has prescribed two separate eye drops, you can apply them at the same time. You should wait a minute or two between applications, simply to make sure the first drops have been distributed by your tear and blinking mechanism, but it is not necessary to wait an hour or more.

Most eye drops will burn slightly in the first few moments after application because of chemical preservatives that have been added to keep the drops stable and sterile. Wait a few moments, do not rub, close your eye if that is more comfortable, and you will find that the sting quickly subsides.

Today most eye drops come prepackaged in plastic bottles with applicator tips—the eyedropper seems to have had its day. The applicator-tip packages are sold under brand names, which makes it difficult for you to buy the medication under its generic name. This is unfortunate, since brand-name medicines tend to be more expensive. If you are concerned about prescription costs, ask your doctor to indicate the generic name so that if it is available as such you can save what might be a considerable amount of money. In most cases, however, you will be unable to find liquid eye medications in this more economical form. Tablet and capsule drugs, which are also prescribed for eye problems, are more widely available under their generic names.

Eyestrain

There is a collection of symptoms that both doctors and patients lump under the term *eyestrain,* which includes tiring of the eyes when reading, a tight or pulling sensation around the brows, and a general feeling of fatigue centered in the eyes. Although the cause can certainly be some sort of eye-muscle imbalance or may indicate a need for glasses, frequently an individual has simply reached his or her threshold of eye use and is ready for a rest. Your doctor can rule out muscle imbalance or the need for glasses, or if he finds that one of these is the cause, can treat it. This is another case in which I strongly advise against seeking an optometrist's counsel—all too often people complaining of eyestrain are sold a weak pair of reading glasses, which is absolutely unnecessary.

Eyestrain is frustrating, but it is important for you to realize that your eyes have limits. If you run around the block for twenty minutes and your legs get tired, this does not mean there is something wrong with your legs and you need crutches; it simply means that you should take a short rest before running again. Your neighbor may be able to run for forty minutes without tiring, and your spouse may be able to manage only ten. The point at which one's eyes get tired is also an individual matter. By and large, you must live with this, but there are some common-sense measures you can take to maximize the amount of time you can read or use your eyes intensely before tiring. Inadequate sleep, tension, poor diet, and poor general health will all contribute to eye fatigue. Take care of yourself, mind and body, and you will probably experience less eyestrain.

And when you do feel your eyes getting tired, take a break. Close your eyes for a few minutes or gaze out the window or even at the ceiling. After this rest you will be able to resume reading with greater comfort.

Proper Lighting

Many people consider proper lighting to be an essential element in preventive eye care. Mothers have been haranguing children for decades, if not centuries, not to read in poor light. Their intentions are admirable, but the fact is that you cannot harm your eyes by reading or doing any kind of seeing in less than "good" light.

This does not mean you should read in the dark. There is a proper level and location of light for reading that will help you see better and cut down on eyestrain. But this is a matter of comfort, not medical eye health.

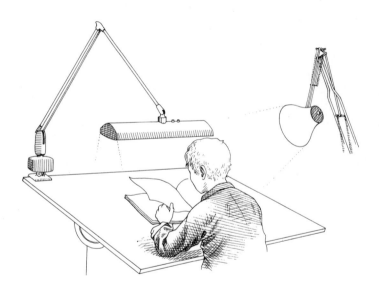

Two proper lighting arrangements: shielded front lighting and over-the-shoulder back lighting.

The best light setup for reading is when the light source is located behind you—it can be behind either or both shoulders —and is directed onto the page. In this way, what you are looking at will be lighted, but you will not have light glaring into your eyes, as you would if the light source were in front of you. Within reasonable bounds, the brighter the light, the better, since it will provide more contrast and make the words on the page easier to read. Obviously, a light so bright that its glare bothers your eyes is too bright.

The second-best setup is with a shielded light in front of you that is aimed at the page. The small high-intensity lamps are the most common example of this. Their effect is essentially the same as the first setup since the shield keeps the light from your eyes and focuses it on the page.

You should choose the sort of lamp and light that provides you with the most comfortable light. Whether it is fluorescent or incandescent light makes no difference. Many people find fluorescent light harsh. If you do, use an incandescent bulb. Neither is better or worse for your eyes.

It is similarly unimportant how the rest of the room is lighted. There is no advantage to a darkened room, leaving you to read in a pool of light, nor is it any better to read in a well-lighted room.

When you can control the manner and type of lighting, arrange it so it is comfortable to you. If it is beyond your control, do not worry that you are injuring your eyesight or eye health. You are not.

Eye Cosmetics

Since our eyes are among the most expressive and psychologically important parts of our bodies, we should feel as good about them as we possibly can. If this includes wearing eye cosmetics, there is no reason you should not do so. There is nothing inherently harmful in using eye cosmetics, but since some people experience difficulty, let's talk about the best and safest way to use them.

Some people have allergic reactions to eye cosmetics. If you are one of these, do investigate the various lines that advertise themselves as hypoallergenic. Of course, any given individual could have an allergic reaction even to these, since none can claim to be 100 percent free of all allergy-causing ingredients. Buy a small amount of one or, better yet, get a free sample if available and give it a try. If you do have an allergic response, discontinue using the cosmetic immediately. In all likelihood, your reaction will subside within a day or two.

Once you have found a product that works for you, you can feel free to use it whenever and as often as you like. Properly used, eye cosmetics are not harmful to your eyes. Those of you who have no allergic problem should buy whatever cosmetics appeal to you.

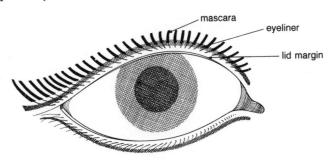

The safe way to apply eye makeup.
Apply mascara to only the outer two-thirds of the eyelashes. Apply eyeliner outside the lashline, keeping it away from the lid margin.

But what is the proper way to apply cosmetics? How can you avoid irritating your eyes? The object is to keep the cosmetics out of the eye itself. For example, when you apply mascara, you should do so on the outer two thirds of your lashes only. Do not begin at the roots, because the mascara can run into your eye or, once it has dried, flake off and fall into your eye. Care should also be taken with the application of eye liner, whether of the liquid or pencil/crayon type. Never apply eye liner to the inner eyelid margins. Instead, put it on just above the eyelashes of the upper lid and just below the eyelashes of the lower lid. If this urging means that you must revise your eye make-up design, forgive me, but I have seen too many patients who lined their eyes instead of their eyelids and ended up with major eye complaints. Keeping just to the outside of the lashes will minimize the chance of poking the pencil or applicator in your eye and having the liner run or flake off into your eye.

When you apply eye liner, do it gently, being careful not to bear down hard on your eyelid. And take care not to poke a pencil or applicator into your eye. If you use the crayon-type liner, you may find it easier to warm the point for a moment with your fingertips to soften it and make application easier and smoother.

Eye shadow or other creams intended for application to the lids should also be kept out of your eyes. Begin a small distance away from the margins and apply gently.

What about false eyelashes? If you feel naked without false eyelashes, you can certainly wear them, but some caution is advised. Regardless of the type you use, you will be working with glue. Be extremely careful not to get any of this in your eyes, and discontinue using it if you have an allergic reaction to the adhesive.

I would also like to recommend against the so-called permanent false eyelashes, the sort that can be left on for several months. There is a danger with these that the normal secretions of your eyelid glands will be blocked, and infection—sties and chalazions among them—can result. At least with the sort of eyelash that you remove each night you will be able to clean your lids and permit gland secretion to proceed normally while you sleep. My personal response to false eyelashes is that they are not terribly attractive and hardly worth the bother.

Another cosmetic treatment I am less than enthusiastic about is eyelash curling. If you must curl your lashes, begin away from the edge of your eyelids so you do not pinch the sensitive skin. And do not pull or twist the curler once your eyelashes are clamped into it. Other than that, you are on your own if curly lashes are your ideal.

How you remove eye make-up is a matter of personal preference. You can wash with soap and water or apply a cleansing cream or oil, or use one of the premoistened, oily pads advertised especially for make-up removal. Whichever you choose, keep your eyes closed when you do this, take care not to let any make-up or remover run into your eyes, and do not bear down hard or rub. The gentlest touch should do the job and avoid irritation of your eyes.

Perhaps the most important thing you can do for your eyes is to take them to an eye doctor for regular checkups. In the next chapter we will discuss the routine eye examination which all of us should have regularly as part of a total program of preventive medical care. But beyond that, and a reasonable respect for your eyes, which includes not poking things into them, removing foreign bodies properly, and responding quickly and intelligently to real eye emergencies, your eyes can largely take care of themselves. It is true they are delicate and vital mechanisms, but nature has provided them with extremely effective armor and defense systems. If you do not abuse those natural defenses, you will in all likelihood find that your eyes will serve you long and well.

11

The Routine
Eye Examination

MYTH: If you need to see an eye doctor, it does not matter whether you go to an ophthalmologist, an oculist, an optometrist, or an optician, since they are all more or less the same.
FACT: They are not all the same, and it matters greatly which you consult about your medical eye health. An optician is not an eye doctor at all; he is a technician who makes corrective lenses according to a doctor's prescription. An optometrist is a doctor of optometry, not an M.D. He is neither trained nor licensed to treat medical eye problems, though he can prescribe corrective lenses. An ophthalmologist is a medical eye doctor, trained and licensed to prescribe corrective lenses, treat diseases of the eye, perform eye surgery, and give complete medical eye care. Oculist is another name for ophthalmologist.

MYTH: If you do not wear glasses, you do not have to have your eyes checked periodically.
FACT: Everyone should visit an eye doctor for regular eye checkups. People who do not wear corrective lenses can wait longer between appointments, but preventive medical examinations of total eye health, particularly past age forty, are as important for people without optical errors as they are for those who have them.

MYTH: A dilated pupil will cause your vision to be blurred.
FACT: Many people believe this is true, since one type of drops an eye doctor may use during the examination to dilate the pupils also causes blurred vision. In fact, this temporary effect is caused by a component of these drops that paralyzes the near-focusing muscles. The loss of focusing ability, not the dilation of the pupils, is what causes the blur.

There is an air of mystery about an eye examination that is absent from other periodic health checkups. When your physician thumps your knee with a rubber mallet, you know that he is checking your reflexes. You might not completely understand the mechanism of reflexes or know what a good or bad reflex indicates, but you do know that your reflex is what is being tested. When your dentist pokes around your mouth with a small metal probe, you know he is looking for cavities. But what is your ophthalmologist doing when he puts your chin on a platform in front of an instrument that looks like something out of NASA Mission Control? And why is he shining a beam into your eyes? Why is he asking you how many dots you see or whether a particular object is above or below a particular line? Why is he putting drops in your eyes, and why do they make it impossible for you to read a magazine while you wait for the drops to take effect? How, after all, can someone else tell what you see and whether or not you see it properly?

It *is* mysterious, and it may often seem as if you were playing children's games with your eye doctor, but the fact is that ophthalmology is a very exact science and all those strange instruments and tests and drops are specialized tools with which your doctor can discover essential diagnostic facts about the state of your vision. Still, why shouldn't you know what it's all about? It will not make you an ophthalmologist or help you pass your doctor's tests, but it may make it easier for you to undergo the all-important eye examination and feel more comfortable, since you will have a general idea of what is going on.

Let's take a step-by-step trip through a basic routine eye examination and find out what the various tests and instruments are, how they work, and what the findings mean. We will be concerned here only with the routine preventive eye test, not with any abnormalities. If the test results indicated a problem, your eye doctor would do further tests and/or take therapeutic measures, but I will focus simply on the basic screening tests I perform on my eye patients. Although every doctor has a somewhat different routine, the similarities are greater than the differences.

Who Should Have an Eye Examination?

Everyone! Even if you do not wear glasses, even if you see well enough, a periodic eye checkup is an important part of your total program of preventive medicine—just like regular physical exams and visits to the dentist. As we grow older, our eyes and our visual needs change, and certain disorders may develop. Some, like glaucoma and cataracts, are specific to the eye; others, like certain circulatory problems, often show up in the eye and can be diagnosed by your eye doctor. And, as you know, in early childhood the routine eye exam is of particularly great importance. Everyone's first visit to an eye doctor should occur at age three, while it is still possible to correct problems that might interfere with visual development. If all is well, the next visit should be the six-year-old pre-school exam. If you wear corrective lenses, you should have your eyes examined about once every eighteen months after that. If you do not need correction, once every two or three years is sufficient until you reach your forties. At that time you should increase your visits to once a year, regardless of whether or not you wear glasses or contact lenses, since this is the age when glaucoma is most likely to develop.

This is the general schedule for periodic preventive eye examinations, though variations may occur from one doctor to another. Your own eye doctor will let you know when you should come in for your next visit.

Total medical eye care can be provided only by an *ophthalmologist,* a specialist in eye medicine. *Optometrists,* who are not medical doctors, can prescribe glasses or contact lenses, but they cannot treat eye diseases. In general, the eye care they give is quite limited and, even in the area of prescription of corrective lenses, they are not trained or, in most states, licensed to use any kind of eye drops, including those that are extremely helpful to your doctor in determining your prescription with complete accuracy.

If you are concerned about the cost of eye care and feel that a private ophthalmological examination is beyond your budget, I recommend that you visit an eye clinic at a nearby teaching hospital. Although there are inconveniences involved, you will be treated by doctors who are getting training in the specialty of ophthalmology and are working under supervision. If there is no such clinic available in your area or if you are reluctant to avail yourself of its services, optometrists are a much less desirable option. I would recommend making this choice, if absolutely necessary, only between the ages of twenty and forty, the period in your life when your eye health is least likely to be complicated, require medical attention, or undergo change. It is generally past the time when childhood eye problems and optical changes occur and before the age when medical eye exams are of greatest value.

One of the reasons people go to optometrists rather than ophthalmologists is that they are attracted by the offer of free eye examinations. What they should understand is that the reason behind this is that optometrists are in the business of selling glasses. In view of the cost of glasses today, a visit to an optometrist can be very expensive indeed. Although there are certainly many ethical optometrists, too many customers are sold glasses that are unnecessary, or they are told they need a change of prescription when they absolutely do not. An example of this that I see in my office every day is the under-forty nearsighted person who has been sold two separate pairs of glasses, one for reading and the other for distance vision. If you understand the optical errors, you will immediately recognize this as a ruse to sell an extra pair of glasses, since virtually no person under forty, who therefore still has near-focusing ability, needs different glasses for near and distance vision. Once the optical error of myopia is corrected with a single pair of glasses, the image comes to focus on the retina, regardless of the distance of the object being viewed. I have also often seen patients who have been told they need their prescriptions changed slightly—and while they are ordering new corrective lenses, why not get new frames?—when a minimal change in their optics does not really require a new pair of glasses. Optometrists are also very fond of prescribing very weak reading glasses, just this side of window glass, for customers under forty who complain of mild eyestrain and headaches on reading.

These three common practices, and various others, seem to me a deplorable shortchanging of the public. That is why I believe that if you are eager to save money on eye care and are between the ages of twenty and forty, you might do better not to have your eyes checked at all. You will surely save money and, in this period of your life, you will be taking the least risk of neglecting the health of your eyes. This does not, of course, apply to people under twenty and over forty, people with medical eye conditions, or those who wear glasses and experience any sort of problem or discomfort with them. These people should see an ophthalmologist; an optometrist is a poor second choice for vision problems, and, if you have a medical eye problem, he cannot treat it at all.

If you do not have a regular ophthalmologist and wonder how to find a good one, you can ask your family doctor to recommend one, or, if your home is located within a reasonable distance of a teaching hospital, you can call the department of ophthalmology there and ask them to give you the name of one or more staff ophthalmologists who have private practices in your area. A third possibility is to call or write the national headquarters of the American Academy of Ophthalmology (see Useful Addresses, page 209) and ask them to send you a directory of member eye doctors in your area. All members of this national academic organization have been granted board certification in the specialty of ophthalmology. Although neither board certification nor affiliation with a teaching hospital is an absolute guarantee of quality, both are good indications of professional expertise.

When you make a first appointment with an eye doctor, do not be afraid to ask about the fee schedule, and if it seems too expensive, do not hesitate to shop around. And when you make your first visit, feel free to decide for yourself whether or not you feel comfortable with that particular eye doctor. Medical expertise is certainly a major quality to look for in a doctor, but a good rapport and feeling of trust are equally important, and you should not be treated by a doctor you do not like.

In the Consultation Room

Examinations in my office begin with a patient history. In order to treat you as an individual, I need to know certain details about the general state of your health, childhood diseases, family medical history, and allergies to drugs or other substances. You have undoubtedly been asked a similar series of questions by your physician, but you may wonder why an eye doctor needs to know these things. For one thing, many eye problems may be related to hereditary factors, so your family eye and medical history will tell me whether or not there are any special signs I should watch out for. For another thing, I may need to prescribe medication, and I want to be sure I do not choose one you may be allergic to. And, finally, your eye health is related in many ways to your total physical health, and the more I know about that the better I am able to look after your eyes.

After obtaining this general information, I will ask a series of questions that zero in on your eyes. Have you had any eye ailments? Do you wear glasses? If so, how long have you worn them? How old is your current pair of glasses? How do you wear your glasses—for reading only, for seeing far away only, or all the time? Do you wear contact lenses? When did you first get them? When were they last checked? Do you experience any particular discomfort or difficulty with them?

Finally, and perhaps most important, I ask if you have any complaints. Although I hope you are visiting me for a periodic checkup, many people go to see their eye doctors only when they have a specific problem, in the same way that many people wait for a toothache before going to the dentist.

In the Examination Room

Up to this point it has all been talk. Now I will take you into the examination room for a series of tests. In some of them I will observe the results (these are called *objective tests*); in others I will ask you to report what you see *(subjective tests);* some of the tests will take place in a darkened room, others in the light; for some tests I will alter your vision in various ways; for others it will be in its natural state. I have a battery of instruments I use, a few of which are simple, everyday objects, while others are very specialized instruments. All the tests are designed to give me specific diagnostic information; they are screening tests, meaning that they tell me how generally healthy your eyes are and help me detect any signs of trouble that I would have to investigate further.

The first thing I do is to test how well you see by asking you to read a Snellen chart—those familiar rows of letters that diminish in size from top to bottom. You will read selected lines of the chart aloud under various circumstances—with (if you wear them) and without corrective lenses, with one eye or the other covered. I will also test your near vision by asking you to read aloud something close up.

If you wear glasses, I will next ask to see them so I can record the prescription. This is important since I will need to know later on whether or not your prescription should be changed. I can ascertain the exact prescription of your glasses by looking at them through a special instrument called a *lensometer,* which "reads" off the full prescription of each lens. If you wear contact lenses, I can also determine their optics with the same instrument.

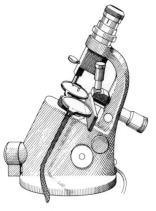

Lensometer.

The next step is a physical examination of your eyes, starting from the outside. I will look at the overall anatomy of your eye area—eyebrows, eyelids, and eyelashes, in addition to the eyeball itself—to see if the area is symmetrical, if there are any growths, scars, redness, excessive tearing, or other signs of abnormality. I will pull down your eyelids and, with a penlight and magnifying lenses, examine the inner surface of the lids.

Then I focus on the eyeball itself. Do your pupils react normally to light? Are your eyeballs reddened? Is there any abnormal pigmentation or growth on the sclera or the conjunctiva? Does the visible portion of your eyeball look normal, and are the various parts in their proper position? If I observe any abnormalities, I will look further for causes, but at this point I am mainly interested in the immediately obvious appearance of your eyes. Later I will make a very minute and detailed examination of both external and internal anatomy.

Next I will check your eye-muscle balance to make sure that your eyes work together normally. The five tests commonly used to determine if the eye muscles are properly coordinated answer three basic questions: Can you move both eyes together in all directions? Can you keep your eyes parallel to each other on all planes? And do you see the same thing with each eye? The form the tests take and the instruments used may vary somewhat from one doctor to another, but they all yield the same information, and although they take some time to explain, I can do each test in less than five seconds.

The first test examines the external movement of your eyes. I will ask you to follow a penlight with both eyes as I move it in a full circle within the periphery of your vision. Meanwhile, I will be watching your eyes to see if you can move them in all directions as far as you should and if both eyes move together.

The four-dot test, one of the basic tests for eye-muscle function.

The second test requires that you wear a special pair of glasses with a red right lens and a green left lens. I will sit facing you and will hold a flashlight that shines four small circular dots at you. Two of the dots are red, one is green, and one is white. The combination of the colored lenses and the color of the various dots makes it possible for each eye to see some dots, but not others. I will ask you to look at the light with both eyes and tell me how many dots you see. You do not have to tell me the color of the dots. There are four possible answers, and each tells me whether or not you are using both eyes together and if not, which you are using. If you say you see four dots, it means you are using both eyes and everything is normal. If you see two dots, you are using your left eye only; three dots tells me you are using your right eye only; and five dots means you have true double vision and are using both eyes but not together.

The next routine test I do checks your vertical muscle balance to be sure that you can hold both eyes on the same level when you look at something. I put a special lens in front of one of your eyes and shine a penlight directly at you while you stare straight ahead and look out of both eyes. You will see a horizontal red line and a white spot of light. I will ask

The cover test, another of the basic tests for eye-muscle function.

whether the spot is above, below, or on the line. If you say it is on the line, this tells me your vertical balance is normal. If you say it is above or below the line, this tells me you have a vertical misalignment. This is how the test works: the eye without the lens sees the spot of light from the penlight; the lens over the other eye turns that spot of light into a red line. Even though both eyes are looking at the penlight, each sees a different thing. When the eyes are aligned, the eye behind the lens will be looking in exactly the same place as the other eye, and the spot and the line will appear to be in the same place. If the eyes are not aligned, the spot will appear to be above or below the line because both eyes will not be looking in the exact same place. Horizontal alignment can also be tested by rotating the lens so the line is horizontal and the spot of light is seen either on the line or to the left or right of it.

I then do an objective test that is very helpful because it can pick up even the subtlest degree of strabismus. I quickly cover and uncover one of your eyes at a time while you stare with both eyes at a penlight straight in front of you. I watch the eye I am testing to see if it jumps when I uncover it. If it remains staring straight ahead, your eyes are properly aligned. If it moves in or out, this is evidence of strabismus, and I have to do further measurements.

If you have normal, three-dimensional vision, this is how the picture of the fly will look to you.

In the next test I check to see if you can use both eyes together to deliver a three-dimensional image to the brain. I will ask you to wear another pair of special glasses and look at a picture, usually of a fly, specially drawn so it will appear three-dimensional with the glasses if you have normal depth perception. If your depth perception is not normally developed, the fly will appear flat, even with the glasses. The principle is exactly the same as that used for 3-D movies or comic books.

At this point in the eye exam, I know how well you can see and whether there is any gross physical evidence of disease or structural abnormality. I also know if you can move your eyes normally, how well they are coordinated, and whether or not you have normal depth perception. Now it is time for that closer, more refined look at the anatomy of your eyes.

Amazing as it may seem, I can, with the help of an instrument and that window into your eye, the pupil, do an internal examination without X-rays or exploratory surgery. The instrument I use is called a *slit lamp,* or *biomicroscope,* and though you may think it resembles a guillotine or some medieval torture device, the exam I do with it is painless. I will ask you to put your chin on the slit lamp's chin rest and look straight ahead while I shine various beams of light into your eyes. The instrument is equipped with a powerful microscope that enlarges the tissues in the parts of the eye I am examining, and the light and lens system works in three ways to permit this minute examination. First, the magnification lets me see things too small to observe with the naked eye. Second, the very bright lights illuminate the darkness inside your eye so I can see where I am looking. Third, the lamp focuses a beam that, along with the lenses, shows me an optically thin section of eye tissue that I can examine much as a pathologist studies a slide under a conventional microscope. In this case, of course, the "slide" consists of living tissue, which is why the instrument is called a bio (living) microscope. This is a particularly valu-

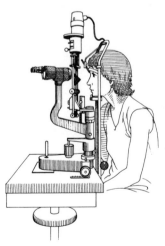

Slit lamp (biomicroscope).

able aid, since removal of eye tissue for study under a microscope cannot be done without seriously injuring the eye.

Through the biomicroscope I take a closer look at the parts of the eye I examined superficially earlier: your eyelids, lashes, surrounding tissues, then the cornea, the sclera, the conjunctiva, and the inside front portion of your eyeball. This examination takes about thirty seconds, and then I move on to a test of your peripheral visual field.

I cover one of your eyes at a time so I can test each eye separately while you focus the other eye on an object (usually my nose) directly in front of you. I will then move a pen back and forth in the outer ranges of the normal visual perimeter and ask you if you see the pen moving. I repeat it for the other eye. If there is a defect in your visual field, you will not see the pen when it is in the defective area. In this case, I will do one of the more detailed visual field tests I told you about in our discussion of glaucoma.

You could, of course, move your eyeball to see the pen, but that would not be evidence of peripheral vision. That is why I keep a close watch to see that your gaze is fixed at my nose throughout the test.

Up to this point, I have done nothing to alter the internal anatomy of your eyes, but now come the drops—the part of the examination that many patients find uncomfortable. Drops are a valuable and essential diagnostic tool, and it will perhaps alleviate much of the anxiety and discomfort you feel if you understand how they work and what their value is.

The general action of the most commonly used drops is twofold: they dilate the pupil so the interior anatomy of the eye can be examined, and they paralyze the near-focusing muscles of the eye so that an accurate refractive reading can be taken. The paralysis is needed in patients under forty or forty-five, since they generally still have focusing ability and my findings would be distorted if their focusing muscles were used. After forty-five or fifty, focusing ability is no longer a factor, so I use drops that dilate only. Whether or not I use paralyzing drops for patients in their low and mid-forties depends on the individual in question.

After I put in the drops, you will be taken to the waiting

room, where you will stay until they take effect; approximately twenty to thirty minutes are needed to get a full dilation with either type. If you are old enough to need dilating drops only, you will now have an opportunity to read the dated magazines usually found there; younger patients, whose vision will be too blurred by the paralyzing action of the drops, may be grateful that reading is impossible.

You may be relieved to know that I, along with many other doctors, do not use drops that paralyze the near-focusing muscles every time I do a routine eye exam. If you are over sixteen and are returning for a regular examination visit, I will probably use drops that dilate only, since I already have an accurate measurement of your optics and it is unlikely that they will have changed materially. Obviously, if my examination or your complaints suggest the need for an examination with the paralyzing drops, they will be used, but it is often unnecessary with return patients. Sometimes very young children, whose focusing muscles are still very strong, and particularly young black children, who have more pigment in their eyes, have a great natural resistance to the paralytic action of the most commonly used drops, so stronger drops have to be used. Because these drops take longer to work, I usually give the parent a prescription for the drops and instructions in their use, so they can be put in at home before a second visit, when the post-drop part of the exam will be done.

Because all these drops take some time to wear off, they sometimes cause discomfort to patients. Light is irritating because the dilated pupils are unable to constrict to exclude excessive light (sunglasses will help alleviate this problem), and vision is blurred up close when the near-focusing muscles have been paralyzed. The effects of simple dilating drops last about three hours, while the drops that dilate and paralyze remain in effect for approximately twenty-four hours. The stronger drops used for some young children can last for two days or more, depending on the type used. In all cases, the action is temporary and harmless, and although it may sound extreme in the case of the stronger drops, a young child who does not yet read or require sharp near vision for numerous other activities will not be too bothered by the blur.

As soon as you have the appropriate drops in your eyes and they have taken effect, I am ready to perform a last series of tests and examination procedures. The first of these is called the *refraction*, and it is the check for corrective lenses. By now I know many things about your eyes, but only by examining your optical system with a special instrument can I determine whether you are nearsighted, farsighted, astigmatic, or have no optical error at all.

There are two main stages in the refraction, one objective and the other subjective. I use the subjective portion to confirm my objective findings, to provide a double check and refinement. But I hope you realize that even without your own responses to the subjective portion, I can make at least a 95 percent accurate finding, close enough for all practical purposes. That is why very young children or mentally retarded or senile adults can still be fitted with an extremely accurate prescription. Patients who are able to cooperate often feel somewhat insecure about their responses to the subjective portion and fear they might give the "wrong" answer and thereby get the wrong prescription, but this concern is unnecessary since the major part of the findings comes from my objective measurement. Any response you make that contradicts my observations will be double-checked.

I use two instruments for the objective portion. One is a *retinoscope* and the other a *phoroptor*, which contains various lenses that can be used singly or in combination to give all the possible optical corrections. I will shine a light from the retinoscope into your eye and ask you to look at the light while I look into your eye through the retinoscope and phoroptor, changing lenses until I obtain an almost exact measurement of your optics. In this way I can determine what prescription would be required to correct your optical error, if indeed you have one. Each eye is examined separately since the precise optics of each are usually different.

The subjective portion will bring this measurement as close as possible to 100 percent accuracy. This is when I ask you to read the Snellen chart again while I hold up a choice of two lenses over the eye that is reading and ask you which looks clearer. This part of the exam makes some patients nervous

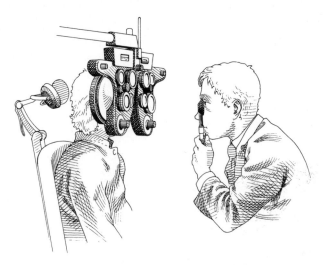

Patient behind phoropter being examined for glasses.
Doctor holds a retinoscope.

since the difference between the two lenses is very slight and they worry that they will choose the wrong one. Do not worry about this—I have my objective findings to guide me and will not let a "wrong" answer pass without double-checking.

We will return to the slit lamp for the next post-drop examination so I can examine the inside back of your eyes. Using the magnification, light, and microscope features of the lamp, I will look through your enlarged pupil. It is, after all, a two-way window, so just as you can see out of it I can look through it. Because your eye is filled with clear structures from front to back, I can examine and then look through the cornea, the anterior chamber, the pupil, the lens, and back to the vitreous fluid. A special lens even gives me a clear view all the way to the retina. If nothing abnormal catches my eye, this check takes about thirty seconds.

The last thing I do with all patients under forty is take a closer look at the inside back of the eye. I will ask you to fix your gaze across a darkened room, and then, holding an instrument that shines a bright light through a lens system, I will get close to you and look through the dilated pupil. The instrument I use is called an *ophthalmoscope,* and it is like a searchlight with which I can actually see your retina, including the macula, optic disc, and the blood vessels of the retina to check if your retina is healthy and if the circulatory system is in good condition.

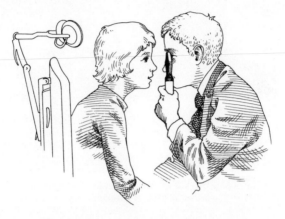

Patient being examined with ophthalmoscope.

If you are over forty or have a family history of glaucoma or show certain signs associated with the disease, I will perform a *tonometry* test to measure your intraocular pressure. I will anesthetize your eyeballs with special drops and, depending on which of two types of instruments I use, will either seat you in a chair tilted so you look at the ceiling while I apply the tonometer to your eyeball or take the measurement with a tonometer attached to the slit lamp. With this second method, a special orange dye is put in your eyes, but the dye is visible only under a special blue light used for this test. Your tears will wash it away completely within a few minutes, so you will not leave my office with weird-looking orange eyes. Each method takes less than five seconds and is completely painless.

The routine examination is now complete. If you wear contact lenses, however, I will check those. I will examine your lenses under the slit lamp, using the same orange dye I use for the tonometry to highlight certain features I want to check on. Essentially, I am looking at how well your lenses fit and seeing if there are any scratches, chips, or warps. Finally, I will check your corrected vision by asking you to read the Snellen chart with your lenses on.

In fifteen to twenty minutes (not counting the time you spent waiting for the dilating drops to take effect) I have

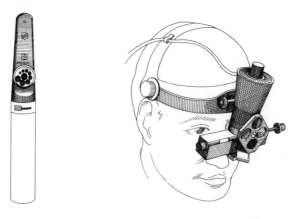

Two types of ophthalmoscopes.

conducted as many as twenty screening tests and diagnostic examinations and have determined the following:
- the medical health of your eyes, externally and internally
- your visual acuity
- your visual field
- the condition and development of your eye-muscle coordination
- whether or not you need optical correction, and what your exact prescription is if you do.

In short, I now have a complete medical picture of your eyes. If the results of all the screening tests are normal, you have medically healthy eyes. If I spotted any abnormalities, I know I may have to do further, more specific testing.

In the Consultation Room

Visits in my office end where they began—in my consultation room. We will sit down together and I will tell you what I found out, explain any important details, and give you whatever prescriptions or instructions are required. If I have determined that you have an optical error that could be corrected with glasses or contact lenses, I will discuss this with you and tell you what I recommend. This does not mean I will insist that you get glasses or contacts regardless of your feelings on the matter. Your visual needs and your judgment about how adequate your uncorrected vision is to those needs are important too. I certainly tell my patients what benefits might be gained with optical correction, but because I know a pair of glasses or contacts left in a bureau drawer will improve vision not one bit, I leave the final decision up to you.

Finally I will tell you when to return for your next appointment and ask if you have any questions. It is at this point that many patients lose the power of speech. Any careful and concerned eye doctor will be more than willing to take the time to explain in greater detail any of his findings and answer any pertinent questions you might have. If it makes you feel more comfortable to be informed, you should by all means ask for information. Your eye doctor will be happy to give it to you.

12

The Layman's
Eye Dictionary

The Layman's Eye Dictionary is a selective collection of definitions that can be used by the reader in several ways. It includes all technical and medical terms mentioned in the text and can be used as a companion to it if further clarification or redefinition is desired. It also includes a number of entries not discussed in the text. Since my intention has been to cover in detail only common eye problems, these entries will serve to introduce the reader to some relatively rare eye conditions, if only briefly. I have by no means attempted to compile a complete eye dictionary; those readers interested in further exploration will have no difficulty finding a variety of adequate lexicons on their library shelves. I have not, for example, included the names of drugs currently used in eye care, nor those of dubious, outmoded, or soon to be outmoded therapeutic procedures.

absolute glaucoma—*See* Glaucoma.

accommodation—The process in which the shape of the lens of the eye adjusts for near vision so that light rays from a near external object are brought to a point of focus on the retina by the accommodated lens.

acute glaucoma—*See* Glaucoma.

allergic conjunctivitis—An inflammation of the conjunctiva in response to an allergy-causing substance.

amblyopia—A failure of normal visual development in an eye without pathological defect and with the anatomical potential for normal vision.

ametropia—Optical error; a condition in which faulty refraction of light rays prevents an image from being brought to focus on the retina, as in the case of myopia, hyperopia, or astigmatism.

aneurysm—An outpouching or enlargement of an artery that can rupture and result in a hemorrhage.

anisocoria—Any inequality in the size of the pupils; can be a normal anatomical variation or a sign of pathology.

anisometropia—A condition in which the optics of one eye differ greatly from those of the other eye.

anterior chamber—The space behind the cornea and in front of the iris that is filled with aqueous fluid.

anterior drainage angle—The area in the anterior chamber at the junction of the cornea and iris at which the aqueous fluid drains from the eye.

aqueous fluid—Also called aqueous humor; the watery fluid that fills the anterior and posterior chambers of the eye; acts to nourish and lubricate the lens and cornea as well as to maintain the eyeball's consistency.

arcus senilis—An opaque, grayish ring at the edge of the cornea that frequently occurs in older people and is the result of fatty deposits in the cornea; has no effect on vision.

arteriosclerosis—A disease of the circulatory system occurring primarily in the elderly; characterized by an inelasticity and thickening of the walls of the arteries, which in turn causes decreased flow of blood. Arteriosclerosis can be diagnosed on examination of the retinal blood vessels; commonly called hardening of the arteries.

artificial tears—A nonmedicinal, nonprescription eye drop that acts as a lubricant and is analogous to thickened natural tear fluid.

astigmatism—An optical error characterized by an unequal curvature in one or more of the eye's refractive surfaces, most commonly the cornea, causing an object to come to focus at two points on the visual axis instead of coming to focus at one point.

Bell's palsy—A peripheral facial paralysis causing the muscles on one side of the face to be completely or partially paralyzed and, thereby, interfering with normal blinking of the eye.

bifocal age group—Refers to that group of individuals of the age (usually mid-forty onward) when reading glasses are needed to correct presbyopia, and, in the presence of another optical error, a second correction is needed for distance vision.

bifocal contact lens—A type of contact lens with two separate optical corrections, one for distance vision and the other for near vision.

bifocal lens—An optical lens having two segments, one for near and one for distance vision.

binocular vision—Vision achieved when both eyes function together; one of the components of normal vision.

biomicroscope—Also called slit lamp; an optical instrument that isolates enlarged and illuminated sections of certain anatomical structures in the eye by utilizing a thin beam of light; particularly useful in examination of the anterior anatomy of the eye as well as the optical media.

blepharitis—A noncontagious infection of the eyelid margin, particularly around the lash area.

blind spot—The small area on the retina at which the optic nerve is attached to the eye. Because it contains no retinal tissue, it is insensitive to light, resulting in a small area of normal visual loss.

bony orbit—The eye socket; the cavity composed of parts of seven bones and containing the eyeball, its surrounding muscles, arteries, veins, and nerves, and surrounding supportive tissue.

cataract—A loss of transparency of any degree and for any reason in the lens of the eye.

cataract surgery—The surgical procedure by which a cataractous lens of the eye is removed.

chalazion—A cyst-like growth near the border of the eyelid that has resulted from an inflammation of a gland at the eyelid margin.

chemical conjunctivitis—An inflammation of the conjunctiva caused by fumes of or contact with chemicals or toxins; sometimes called toxic conjunctivitis.

chorioretinitis—Any inflammation of the choroid and retinal layers of the eye, usually caused by an infection elsewhere in the body.

choroid—A delicate membranous layer of the eye that lies between the sclera and the retina and is continuous with the iris and the ciliary body in front of it; primarily a blood vessel layer.

chronic simple glaucoma—*See* Glaucoma.

cilia—The eyelashes.

ciliary body—A vascular and muscular structure that lies between the choroid layer and the iris; secretes aqueous fluid into the posterior chamber, and its muscles form part of the near-focusing system.

closed angle glaucoma—*See* Glaucoma.

cold sore—Common term for a type of blister, often seen in the mouth and nose area, caused by the herpes simplex virus. *See also* Herpes Simplex.

color blindness—A genetically inherited birth defect that occurs almost exclusively in men and causes loss of normal color perception; the degree of loss varies widely.

concave lens—A lens of which one or both surfaces are curved inward. *Compare* Convex Lens.

cones—Specialized cells of the retina sensitive to color and light intensity; important in light adaptation and color perception.

congenital cataract—A rare type of cataract present at birth. *See also* Cataract.

congenital glaucoma—*See* Glaucoma.

conjunctiva—The mucous membrane covering the exposed front portion of the sclera and continuing to form the lining of the inside of the eyelids.

conjunctivitis—Any inflammation of the conjunctiva. *See also* Allergic, Chemical, and Infectious Conjunctivitis.

constrictor muscles—*See* Dilator Pupillae and Sphincter Pupillae.

contact lenses—A pair of small optical lenses generally made of plastic that correct optical defects inconspicuously by resting directly on the tear layer of the cornea.

convergence—The normal process in which both eyes move inward toward one another to focus on a near object.

convergence reflex—A normal, involuntary reflex in which the visual axes of the two eyes bend toward one another as the near-focusing muscles come into use.

convergent strabismus—*See* Esotropia.

convex lens—A lens of which one or both surfaces are curved outward. *Compare* Concave Lens.

cornea—The curved, transparent membrane forming the front one-sixth of the outer coat of the eyeball; serves primarily as protection and is the outermost refractive surface of the eye.

corneal contact lenses—Contact lenses that are fitted exclusively on the corneal surface; the most commonly used contact lenses today are corneal contact lenses.

corneal dystrophy—Degeneration of the layers of the cornea resulting in noninflammatory lesions that show up as clouding and cause corresponding visual loss.

corneal transplant—A surgical procedure in which a portion of clear donor cornea is used to replace a corresponding portion of the patient's opacified cornea.

corneal ulcer—A defect in the protective outer layer of the cornea, most commonly caused by an infection.

cosmetic contact lens—A type of contact lens used to alter the color of the eyes by providing the iris with an opaque covering that resembles an iris but is of another color.

crossed eyes—A lay term for strabismus. *See* Strabismus.

cryosurgery—Surgery involving the freezing of tissues; used in some cataract extractions, glaucoma operations, and retinal repairs.

cupping of optic disc—An excavated or c-shaped portion of the optic disc; can be a normal anatomical variation or can occur in glaucoma, in which case it may worsen.

cycloplegic drops—Dual-action diagnostic eye drops which temporarily paralyze the near-focusing muscles of the eye as well as dilate the pupil.

dacryoadenitis—Any infection of the lacrimal gland.

dacryocystitis—Any infection of the lacrimal sac.

dacryocystorhinostomy—A surgical procedure in which an

artificial opening is created between the lacrimal sac and the nasal cavity to facilitate tear drainage when scar tissue has blocked the normal drainage pathways.

dark adaptation—The ability of the eye to adjust itself to decreasing illumination. This process occurs in the retina and entails an increase in the number of functioning rods and a decrease in the number of functioning cones. Complete dark adaptation takes about one hour. *Compare* Light Adaptation.

dendritic ulcer—An ulcer of the cornea caused by the herpes simplex virus; has a characteristic appearance resembling the branches of a tree.

deviant eye—Used in reference to strabismus, the eye that is not fixing on the object being viewed but is turned away from the visual axis.

diabetes, diabetes mellitus—A metabolic disorder characterized by an imbalance in blood sugar levels in the body; can have damaging effects on the eye as well as on the rest of the body.

diabetic cataract—A clouding of the lens of the eye caused by diabetes. *See also* Cataract.

diabetic retinopathy—A catchall term used to describe any of the various stages of retinal pathology caused by diabetes, including hemorrhages, thrombi, aneurysms, and scarring of the retinal tissue.

dilator muscles—*See* Dilator Pupillae and Sphincter Pupillae.

dilator pupillae—A muscle situated in the peripheral part of the iris whose contraction causes dilation of the pupil and whose relaxation allows the pupil to contract.

diopter—The unit used to measure the light-bending power of a lens.

diplopia—Double vision; a pathological condition of vision in which a single object is perceived as two.

distance vision—Use of the eyes to view objects in the distance. Theoretically, that distance is infinity; the opthalmological measurement of distance vision is made twenty feet from the eye; in practical terms, one uses distance vision to view anything that is beyond what is considered middle distance. *See also* Middle-Distance Vision.

divergence—The normal movement or spreading apart of the eyes in an outward direction.

divergent strabismus—*See* Exotropia.

double vision—*See* Diplopia.

dyslexia—An inability to read with comprehension at a level indicated by the individual's overall intelligence and/or verbal skills; in medical usage, refers to problems caused by a central (neurological) defect; in common usage, inaccurately used to describe any reading difficulty, such as letter inversion.

ectropion—A condition in which the eyelid margin (usually the lower eyelid) turns outward so that the conjunctiva lining it is exposed.

edema—An excessive accumulation of clear watery fluid in any tissue, resulting in swelling of that tissue.

embolism—The blockage of a blood vessel by a blood clot that has been carried by the bloodstream and become impacted in some portion of the circulatory system, often causing a secondary hemorrhage.

emmetropia—The normal optical condition of the eye that permits rays of light to focus accurately on the retina.

entropion—A constant or intermittent turning inward of the eyelid margin (usually the lower eyelid), causing the eyelashes to rub against the cornea; generally occurs only in the elderly; causes the eye to become red and irritated.

enucleation—A surgical procedure in which the entire eyeball is removed.

epicanthal fold—*See* Epicanthus.

epicanthus—An anatomical feature in which a fold of skin extending from the top of the nose to the inner end of the eyebrow overlaps the inner corner of the eye; in some circumstances can give the appearance of strabismus; tends to recede as the bridge of the nose narrows in the course of early childhood.

epiphora—A condition in which tears overflow onto the cheek due to a narrowing of the tear removal apparatus, or, less commonly, to an excessive secretion of tears.

episcleritis—*See* Scleritis.

esophoria—A tendency of one eye to deviate inwardly when the eyes are at rest, such as when the eyes are closed; to a small degree, can be normal.

esotropia—A type of strabismus in which the deviation of one eye is inward, or convergent, at rest or in use; also called convergent strabismus.

exophoria—A tendency of one eye to deviate outwardly when the eyes are at rest; to a small degree, can be normal; to a larger degree, can cause eye fatigue.

exophthalmos—A forward protrusion of the eyeball frequently caused by an overactive thyroid gland or any type of growth located in the bony orbit.

exotropia—A type of strabismus in which the deviation of one eye is outward, or divergent, at rest or in use; also called divergent strabismus.

external eye muscles—*See* Extraocular Muscles.

extraocular muscles—The six muscles of each eye, which are attached to the outside of the eyeball behind the conjunctival cul-de-sac and fastened by way of a system of connective tissue to the bony orbit, function to move the eye in its various directions to permit full 360-degree movement of the eyes; consist of the medial rectus, lateral rectus, superior rectus, inferior rectus, superior oblique, and inferior oblique muscles.

eye bank—A medical facility that keeps eyes removed from deceased donors to be used in corneal transplantation.

eyebrow—The crescent-shaped line of hair at the upper edge of the orbit.

eyelid—One of the two movable folds of skin lined on the inside with conjunctival membrane and contiguous on the outside with the skin of the face; functions to clean, lubricate, and protect the eyeball.

eyelid tic—*See* Fibrillation.

farsighted astigmatism—A type of astigmatism in which both points of focus have not yet been reached when light rays arrive at the retina.

farsightedness—*See* Hyperopia.

fibrillation—A twitching of one or both eyelids, usually the result of tension or anxiety; commonly called eyelid tic.

fixing eye—Used in reference to strabismus, the eye that is looking straight along the visual axis at the object being viewed.

floaters—Also called vitreous floaters; common term used to describe an aggregate of cells or protein in the vitreous fluid seen as small, moving black shapes, especially visible against a light background; usually of no pathological significance.

fluorescein angiography—A diagnostic procedure utilizing a special dye and photographic technique to examine the fundus of the eye.

focal length—The distance between a refractive surface and the point at which the light rays meet.

focus—The point at which light rays meet after passing through a refractive surface.

focusing ability—*See* Accommodation.

focusing muscles—Common term used to denote the anatomical system used in the process of accommodation, consisting of the ciliary body muscles, the zonule, and the lens itself.

fovea centralis—The small, normal anatomical depression located in the center of the macula.

fundus—The back portion of the interior of the eyeball consisting primarily of the retina, the optic disc, and the retinal blood vessels; can be viewed and examined with the ophthalmoscope.

fusion—The process by which the two disparate images seen by the two eyes are blended into one image to produce binocular vision.

glaucoma—An eye disease characterized by increased pressure within the eyeball.

absolute glaucoma—A final stage of untreated glaucoma generally characterized by blindness and eyes that are painful and chronically inflamed.

acute glaucoma—A glaucoma characterized by a sudden and severe increase in intraocular pressure causing severe pain, redness, and visual blur.

closed angle glaucoma—Another name for acute glaucoma.

congenital glaucoma—A type of glaucoma that occurs at birth or in infancy due to birth defects in the eye.

chronic glaucoma—The common type of asymptomatic glaucoma in which there is a slight and painless increase in intraocular pressure. (*See also* Simple Glaucoma and Secondary Glaucoma.)

glaucoma suspect—A diagnosis in which, in the absence of clinical glaucoma, some tests point to a tendency toward glaucoma warranting more frequent than normal observation.

low tension glaucoma—A very rare type of glaucoma in which levels of intraocular pressure within the normal range cause glaucomatous changes.

open angle glaucoma—Another name for chronic glaucoma.

secondary glaucoma—An increase in intraocular pressure caused by a known problem or medical disorder within the eyeball; for example, a previous eye infection. Both chronic and acute glaucoma can result from a secondary cause. *Compare with* Simple Glaucoma.

simple glaucoma—A type of chronic glaucoma in which the increase in intraocular pressure cannot be attributed to a known underlying cause. *Compare with* Secondary Glaucoma.

gonioscope—An instrument equipped with prisms that can be placed on the cornea and will permit a clear view of the anterior drainage angle; of importance in the classification and treatment of glaucoma.

graduated focal-length lens—A type of corrective glasses in which, from the top distance to the bottom reading correction, there is a range of intermediate focal distances; a new type of lens generally considered of limited use.

heat cautery—A surgical technique that uses heat applied through a small probelike instrument to cause scar tissue to form where desired; primarily used in retinal detachment repair.

hemorrhage—Bleeding; the seepage of blood from any type of blood vessel.

herpes simplex—A virus that can infect the eye and sometimes can cause dendritic ulcer.

herpes zoster—A virus that can in rare cases infect the cornea or eyelids; the virus responsible for shingles.

heterophoria—A condition in which there is an inward, outward, or upward tendency when the eyes are at rest; esophoria, exophoria or hyperphoria.

heterotropia—A condition in which there is an inward, outward, or upward deviation of one eye at rest or in use; that is, esotropia, exotropia, or hypertropia; encompasses all strabismus.

hordeolum—An infection of a gland of the eyelid margin; the common sty.

hydrophilic contact lens—The so-called soft contact lens; a type of contact lens made of plastic that is capable of absorbing and retaining water.

hyperopia—Commonly called farsightedness; the optical error in which an image has not yet come to a point of focus when it reaches the retina; also called hypermetropia.

hyperphoria—A tendency of one eye to deviate upward when the eyes are at rest; a type of vertical eye muscle imbalance that can cause eye fatigue.

hypertension—An increase above normal blood pressure levels; can affect the eyes as well as the body as a whole.

hypertensive retinopathy—A catchall term used to describe any of the various stages of retinal pathology caused by hypertension, including hemorrhages, thrombi, and papilledema.

hypertropia—A type of strabismus in which there is an upward deviation of one eye in relation to the other when the eyes are at rest and in use.

hyphema—A hemorrhage from the iris into the anterior chamber of the eye, frequently as a result of trauma.

hypopyon—The presence of a pus-like fluid in the anterior chamber of the eye secondary to an infection elsewhere in the eye.

image size variation—Phenomenon caused by the correction of anisometropia with eyeglasses, with the result that one eye sees an image of a different size than that seen by the other eye.

index of refraction—A measurement of the refractive power of a surface; how much that surface bends a ray of light that passes through it.

infectious conjunctivitis—An inflammation of the conjunctiva caused by a bacterial or a viral infection.

interstitial keratitis—An inflammation of the deeper layers of the cornea, causing clouding of the cornea; most frequently caused by syphilis or tuberculosis.

intraocular pressure—The degree of firmness of the eyeball as controlled by secretion and drainage of aqueous fluid.

iridectomy—A surgical procedure in which a portion of the iris is removed; frequently performed in glaucoma operations to facilitate drainage of aqueous fluid and as part of cataract extractions to reduce the risk of secondary glaucoma.

iridencleisis—A surgical procedure in which a section of the iris is pulled through the sclera to create a new drainage outlet for aqueous fluid; performed in certain cases of glaucoma.

iris—A disclike diaphragm that is continuous in back with the ciliary body and is perforated in the center by the pupil; composed of vascular and muscular tissue, the latter controlling the size of the pupil. The color of the iris determines the color of an individual's eyes.

iritis—Any inflammation of the iris.

irregular astigmatism—An unusual type of astigmatism in which the unequal curvature of the cornea causes an image to be focused at more than two points along the visual axis.

keratitis—Any inflammation of the cornea.

keratoconjunctivitis—Any combined inflammation of the cornea and conjunctiva.

keratoconus—A cone-shaped protrusion of the center of the cornea caused by a progressive thinning of the corneal tissue layers because of a structural weakness; its effects are high irregular astigmatism and eventual rupture of the cornea if untreated.

keratometer—An optical instrument used to measure the curvature of the cornea; of particular use in the fitting of contact lenses.

lacrimal apparatus—The system responsible for the formation, secretion, and drainage of tears. Includes:

lacrimal canal—A small channel along which tears drain from the punctum to the lacrimal sac.

lacrimal duct—The passageway by which tears are drained from the lacrimal sac into the back of the nose and, from there, into the throat.

lacrimal fluid—Tears.

lacrimal gland—The gland that manufactures the tears; located in the upper, outer wall of the orbit.

lacrimal lake—The collection of tears in the conjunctival sac where the lower eyelid conjunctiva and scleral conjunctiva meet.

lacrimal punctum—A tiny elevation with a central hole located on the lower eyelid margin toward the nose; its function is to siphon off the tears.

lacrimal sac—A small saclike pouch between the lacrimal canal and the lacrimal duct; forms part of the tear drainage system.

laser beam—A powerful, concentrated beam of light that can generate a great deal of heat in a tiny, specified area; can be used in retinal surgery.

lazy eye—Lay term for amblyopia. *See* Amblyopia.

legal blindness—Vision that cannot be corrected to better than 20/200, or the presence of extreme tunnel vision.

lens (optical)—A transparent substance, usually made of glass or plastic, with two opposed surfaces, used to bend light and thereby change the point at which light rays focus, such as is used in the correction of optical errors.

lens (crystalline)—A transparent, flexible body, convex on both surfaces and lying directly behind the iris of the eye; serves to focus the rays of light on the retina.

lensometer—An optical instrument used to determine the exact prescription of a pair of eyeglasses.

light adaptation—The ability of the eye to adjust to brighter illumination. This process occurs in the retina and entails an increase in the number of functioning cones and a decrease in the number of functioning rods. Complete light adaptation takes about one hour. *Compare* Dark Adaptation.

low tension glaucoma—*See* Glaucoma.

macula—The specialized central area of the retina, which is responsible for sharp central vision; surrounds the fovea centralis.

malignant melanoma—A rare cancerous tumor originating in the choroid layer inside the eyeball.

manifest refraction—The subjective portion of the refraction in which the subject reports to the examiner which of two lenses provides superior vision; also called manifest.

meibomian gland—One of about thirty or forty small glands near each eyelid that secrete a small amount of fatty lubricant at the eyelid margin to prevent normal amounts of tears from flowing over onto the cheek.

microphthalmos—A condition in which the eyeball is abnormally small; usually a congenital condition.

middle-distance vision—Use of the eyes to view objects in the area beyond arm's reach but within normal conversational distance; not a routinely measured ophthalmological distance, but in practical terms between two and a half and five feet from the eyes. *Compare* Distance Vision and Near Vision.

miosis—Contraction of the pupil.

miotic drops—A type of eye drop causing miosis and used therapeutically in the control of glaucoma.

mixed astigmatism—A type of astigmatism in which one point of focus is in front of the retina and the other is theoretically behind the retina.

myopia—Commonly called nearsightedness; the optical error in which an object comes to a point of focus before it reaches the retina and is thus out of focus on the retina.

nearsighted astigmatism—A type of astigmatism in which both points of focus are in front of the retina.

nearsightedness—*See* Myopia.

near vision—Ophthalmologically, measured at 14 inches (33 centimeters), a normal reading distance; in practical terms, vision used to view objects within arm's reach of the eyes, or up to about two feet. *Compare* Distance Vision and Middle-Distance Vision.

night blindness—Any of several rare eye diseases in which degeneration of specialized retinal tissue causes abnormally poor vision in dim light. *See also* Retinitis Pigmentosa.

nonmedical contact lenses—Contact lenses used primarily for cosmetic and optical purposes, as opposed to those used to treat or control medical eye problems.

normal vision—Two eyes with 20/20 visual acuity and fully developed binocularity.

nystagmus—An involuntary, rhythmical oscillation of the eyeballs in a horizontal, vertical, or rotary direction; may be caused by a central (neurological) problem or a localized eye problem, and can either cause or result from faulty visual development.

objective test—A diagnostic test whose results are observed and determined by the tester, rather than reported by the subject.

ocular motility—The movement of the eyes, singly or together, laterally, vertically, obliquely, divergently, or convergently.

open angle glaucoma—*See* Glaucoma.

ophthalmologist—Also called oculist; a medical doctor trained in the diagnosis and treatment of eye diseases and correction of optical errors.

ophthalmoscope—An instrument equipped with a system of mirrors and lights and with which the fundus of the eye can be examined.

optic atrophy—Any condition in which the fibers of the optic nerve degenerate and finally die, causing corresponding loss of vision; glaucoma is an example of optic atrophy.

optic canal—The bony tunnel through which the optic nerve runs from the back of the bony orbit into the front of the brain cavity.

optic chiasm—An x-shaped area in the brain where the optic nerves of the right and left eyes meet and cross.

optic disc—That portion of the optic nerve at the point of entrance into the back of the eye; corresponds to the location of the blind spot.

optic nerve—The nerve of sight; the collection of specialized nerve fibers derived from the retina which unite and send visual impulses to the brain.

optic neuritis—Any inflammation of the optic nerve.

optic tract—A specialized group of nerve fibers that begins in the optic chiasm and carries the visual impulses to the brain.

optical error—*See* Ametropia.

optical media—The transparent structures of the optical system through which light passes to focus an image on the retina, including the cornea, aqueous fluid, lens, and vitreous fluid.

optician—A specialist who makes glasses in accordance with a doctor's prescription.

optometrist—A doctor of optometry who is trained to test the eyes for nonmedical defects of vision in order to prescribe and dispense corrective lenses.

ora serrata—A serrated edge located just behind the ciliary body that delineates the front ring of the retina and marks the limits of the retina's perceiving portion.

orbit—*See* Bony Orbit.

orthophoria—The condition in which the two eyes are in perfect parallel alignment whether at rest or in use; a completely normal balance of the eye muscle functions.

orthopic exercises—Training with eye exercises used effectively only occasionally and only in very young children in an attempt to improve eye muscle balance.

oxygen-permeable contact lens—Also called gas-permeable contact lens; a type of contact lens that permits the exchange of oxygen through the lens material itself.

papilledema—A swelling of the optic disc usually caused by interference with the blood circulation of the optic nerve through increased pressure from the brain cavity, by an inflammation of the optic nerve, or by hypertension.

perimeter—An instrument with which the peripheral visual fields as well as the blind spots of the eyes are tested.

peripheral vision—Side vision; visual perception to all sides of the central object being viewed; that which is not central vision.

peripheral visual fields—*See* Visual Fields.

phako-emulsification—A cataract extraction technique in which the lens of the eye is sucked out through a needlelike probe.

phoroptor—An instrument equipped with a broad selection of optical lenses, which, used singly or in combination, provide all possible optical corrections; used primarily in testing the optics of the eye and determining the prescription for corrective lenses.

pingueculum—A fatty deposit that appears as a small, raised, yellowish area on the horizontal midline of the sclera on either side of the cornea; is nonpathological but can enlarge and become reddened; has a greater tendency to appear in older people; does not interfere with vision.

pink eye—*See* Conjunctivitis.

posterior chamber—The space behind the iris and in front of the lens which is filled with aqueous fluid; aqueous is secreted into the posterior chamber and flows from there into the anterior chamber.

presbyopia—The normal physiological change in the ability to focus on near objects which occurs throughout life, normally becomes symptomatic in the forties and results in total loss of ability to focus on near objects around age sixty; normally requires correction with reading glasses in the mid-forties; results from the increased inelasticity in the crystalline lens of the eye.

prism—A transparent solid with a triangular base used to disperse light into a color spectrum or to deflect rays of light toward the base of the triangle; of use in glasses to correct high degrees of extraocular muscle imbalance.

progressive myopia—A lay term most commonly used in reference to the tendency of myopia to worsen during the growth period.

provocative tests for glaucoma—A series of diagnostic tests that create a stress on the circulation of aqueous fluid within the eyeball and will increase intraocular pressure in an eye not healthy enough to withstand the stress. Used to assist in the diagnosis of glaucoma, the tests involve in part drinking large quantities of fluid and sitting in a dark room for a prolonged period of time in conjunction with a monitoring of intraocular pressure.

pterygium—A triangular growth of thickened conjunctival tissue usually extending from the portion of the conjunctiva nearest the nose to the border of the cornea or beyond, with its apex pointing toward the center of the cornea.

ptosis—A drooping of the upper eyelid due to a congenital fault or to an acquired paralysis of the muscle that lifts the eyelid.

pupil—The circular hole in the center of the iris.

refraction—The ophthalmological examination by which the nature and degree of the optical errors present in the eye and the correction of these optical errors are determined. Or the bending of a ray of light as it passes from air into a medium of greater optical density.

refractive ability—The ability of an optical surface to bend light. *See also* Index of Refraction.

refractive surface—Any surface having the ability to bend light. *See also* Index of Refraction.

retina—The thin, delicate, transparent sheet of nervous tissue that lines the back two-thirds of the eyeball; functions as the receptor of visual stimuli, which it transmits to the brain via the optic nerve.

retinal detachment—A separation of the retina from the choroid.

retinitis—Any inflammation of the retina.

retinitis pigmentosa—A rare disease characterized by chronic and progressive degeneration of the retinal pigmentation; hereditary in nature, usually results in little or no vision by middle age; night blindness is an early symptom.

retinitis proliferans—A pathological condition characterized by a series of retinal hemorrhages, which in turn cause scar tissue to form on the retina and in the vitreous fluid, often resulting in severe visual loss and possible detachment of the retina; most commonly seen in diabetes.

retinoblastoma—A rare malignancy originating in the retina, often occurring in both eyes, usually in very young children.

retinoscope—An optical instrument that, through a system of mirrors and lights, is used to detect refractive errors in the eye.

retrobubular neuritis—Any inflammation of the portion of the optic nerve lying behind the eyeball; generally accompanied by poor vision and characteristically seen in connection with multiple sclerosis.

retrolental fibroplasia—An abnormal growth of scar tissue in the vitreous fluid accompanied by abnormal blood vessel growth and retinal detachment; caused by the formerly common practice of administering high concentrations of oxygen to premature infants.

rhodopsin—Also called visual purple; the pigment found in the rods of the retina, chemically important to dark adaptation.

rods—Specialized cells of the retina sensitive to low intensities of light and important in dark adaptation.

sclera—The curved, opaque, protective white layer forming the back five-sixths of the outer coat of the eyeball; the white of the eye.

scleral buckling—A surgical eye procedure sometimes used in retinal detachment repair; the sclera is indented to create a groove, which in turn shortens the eyeball.

scleral contact lens—A type of contact lens no longer in use which covered part of the sclera as well as the entire cornea.

scleritis—Any inflammation of the sclera.

scotoma—An abnormal area of varying size and shape within the visual field in which there is a partial or complete loss of vision.

secondary cataract—A type of cataract caused by a known condition of the eye or general health other than simply advancing age. *See also* Cataract.

senile cataract—The common type of cataract caused by advancing age. *See also* Cataract.

single-vision lens—An optical lens having one focal length; a lens in which there is only one optical correction.

slit lamp—*See* Biomicroscope.

Snellen chart—The standard eye chart employed in the testing of distance visual acuity.

sphincter pupillae—A circular muscle of the iris lying close to the pupillary margin; its contraction causes constriction of the pupil and its relaxation causes dilation of the pupil. *Compare* Dilator Pupillae.

stereopsis—The visual perception of objects as three-dimensional rather than as all in one plane; depth perception.

strabismus—A constant failure of the eyes to maintain parallel visual axes; commonly called crossed eyes.

sty—*See* Hordeolum.

subconjunctival hemorrhage—Bleeding from a burst blood vessel in between the conjunctiva and sclera, resulting in a very red but painless eye.

subjective test—A diagnostic test in which the results are reported by the test subject rather than observed by the tester.

subluxation of the crystalline lens—A dislocation of the lens of the eye due to a rupture in the zonular attachment of the lens. The position the lens assumes depends on the location and extent of the rupture.

sympathetic ophthalmia—An inflammation of the entire uveal tract of one eye invariably caused by a perforating wound that involves the uveal tissue of the other eye.

synechia—Scar tissue extending from the iris, most commonly following an infection inside the eye and possibly causing secondary glaucoma or cataracts. Called anterior synechia if the scar tissue extends forward to the cornea; posterior synechia if the scar tissue extends backward to the lens.

tangent screen—A flat, blackboardlike chart used in one type of test of the peripheral visual fields and blind spots of the eyes.

thrombosis—The clotting of blood in any part of the circulatory system blocking the passage of blood through the involved blood vessel.

tonography—A test that can be performed as part of a glaucoma diagnostic workup, which monitors electronically the efficiency of the secretion/drainage mechanism inside the eye.

tonometer—An instrument used to measure intraocular pressure; a principal part of diagnostic screening for glaucoma.

tonometry—The diagnostic test performed with a tonometer.

toxic conjunctivitis—*See* Chemical Conjunctivitis.

trabeculectomy—A surgical eye procedure sometimes performed for glaucoma in which the drainage of aqueous fluid through the already existing drainage system is improved.

trachoma—A contagious external eye infection; if untreated, may cause scarring of the cornea; a common cause of blindness in underdeveloped nations.

traumatic cataract—A type of cataract caused by external trauma. *See also* Cataract.

trichiasis—An inversion of one or more of the eyelashes causing friction that results in irritation of the cornea or conjunctiva.

trifocal lens—An eyeglass with three different optical portions, one for near, one for distance, and one for middle-distance vision.

20/20 vision—The expression of visual acuity indicating that the test subject can see at twenty feet what a normal-seeing person sees at twenty feet; one of the components of normal vision; also expressed as 6/6 to represent six meters rather than twenty feet.

uveal tract—The vascular and muscular middle layer of the eyeball, consisting of the iris, the ciliary body, and the choroid.

uveitis—Any inflammation of the uveal tract.

visual acuity—The sharpness of vision as determined by a comparison with normal optical ability to define certain letters at a given distance, usually twenty feet. *See also* 20/20 Vision.

visual axis—The imaginary line extending from a viewed object to the retina and passing through the air, corrective lenses (if any), the cornea, the aqueous fluid, the pupil, the crystalline lens, and the vitreous fluid.

visual fields—The entire view encompassed by the eye when it is looking in a given direction; the area visible to the eye, including sharp central and peripheral vision; also called peripheral visual fields.

visual purple—*See* Rhodopsin.

vitreous floaters—*See* Floaters.

vitreous fluid—Also called vitreous humor; a transparent, gel-like substance that fills the bulk of the interior of the eyeball; it begins behind the lens and attaches to the retina at the back and on the sides.

xanthelasma—Cholesterol deposits that appear as yellowish raised bumps on the surface of the skin on or near the upper and lower eyelids.

zonule—The transparent membrane by which the crystalline lens is attached to the ciliary body.

Useful Addresses

1. American Academy of Ophthalmology
   15 Second Street SW
   Rochester, MN 55901
   507–288–7444
   To obtain listing of board-certified ophthalmologists in your area.

2. Contact Lens Association of Ophthalmologists
   40 West 77th Street
   New York, NY 10024
   212–595–0778
   A medical organization that will provide you with a list of medical contact lens fitters in your area.

3. Guild of Prescription Opticians
   1250 Connecticut Avenue NW
   Washington, D.C. 20036
   212–659–3620
   An organization that can provide you with a listing of Guild opticians in your area. Such a listing is also available in your classified telephone directory listings.

4. *Directory of Medical Specialists*
   (Published by Marquis—*Who's Who*)
   A book listing board-certified ophthalmologists in your area, updated every two years; available at public libraries.

Index